Alfred Pancoast Boller

Practical Treatise on the Construction of Iron Highway Bridges

For the Use of Town Committees

Alfred Pancoast Boller

Practical Treatise on the Construction of Iron Highway Bridges
For the Use of Town Committees

ISBN/EAN: 9783743465060

Manufactured in Europe, USA, Canada, Australia, Japa

Cover: Foto ©berggeist007 / pixelio.de

Manufactured and distributed by brebook publishing software (www.brebook.com)

Alfred Pancoast Boller

Practical Treatise on the Construction of Iron Highway Bridges

BRIDGE AT PLAINFIELD, N. J. BY THE AUTHOR. 104 FEET SPAN; 25 FEET ROADWAY; 8 FEET SIDEWALKS; 100 POUNDS PER SQUARE FOOT; FACTOR OF 5 FOR SAFETY.

PRACTICAL TREATISE

ON THE

CONSTRUCTION

OF

IRON HIGHWAY BRIDGES.

FOR THE

USE OF TOWN COMMITTEES.

TOGETHER WITH A

SHORT ESSAY UPON THE APPLICATION OF THE PRINCIPLES OF THE LEVER TO A READY ANALYSIS OF THE STRAINS UPON THE MORE CUSTOMARY FORMS OF BEAMS AND TRUSSES.

BY

ALFRED P. BOLLER, A.M.,

CIVIL ENGINEER,

MEMBER OF THE AM. SOC. CIV. ENGINEERS.

FOURTH EDITION.

SECOND THOUSAND.

NEW YORK:
JOHN WILEY AND SONS,
53 East Tenth Street,
1893.

Entered, according to Act of Congress, in the year 1876, by
ALFRED P. BOLLER,
in the Office of the Librarian of Congress, at Washington.

DEDICATION.

TO

TOWN COMMITTEES, SELECTMEN, COUNTY FREEHOLDERS,

AND

OTHER PUBLIC OFFICERS,

TO WHOM IS INTRUSTED THE RESPONSIBILITY OF
ERECTING "IRON BRIDGES," THIS BOOK IS
RESPECTFULLY DEDICATED

BY

THE AUTHOR.

INDEX.

PART I.

GENERAL AND DESCRIPTIVE.

	PAGE
Æsthetical Effect vs. Plain Utility	87
American and Riveted Systems Compared	61
Angle Iron	26
Angle Irons, Section of, how Determined	64
Architecture of Bridge Building	82
Asphalt, Weight of	79
Author's Bridge	*Frontispiece.*
Beam Bridges	74
" " Plank Floor	75
" " Wood Pavement on Buckle Plates	75
" " Telford Pavement on Brick Arches	76
Bearings and Connections, Machine Made	47
Bolster Pieces	71
Braces, Main and Counter	33
" Proportions of	33
Bracing, Horizontal	67
" Sway	67
Brick Work, Weight of	70
Bridge Settings	90–96
" Platforms	74
Bridges, General Rules for Selection of	43
" How Classified	32
" Kinds of	32
" Selection of	43
Buckle Plate for Floor	73
Cast Iron	28
Chords, Strains in	32
" Their Office	32
Cincinnati Bridge Co.'s Bridge	90
Columns, Cast Iron, Formula for their Strength	57
" Table of Breaking Strength for Cast and Wrought Iron	58
" With Square Ends and with Round Ends	56
" Wrought Iron Formula for Strength	57

	PAGE
Compound Riveted Girder	64
Concrete, Weight of	70
Construction, Methods of	44
Counter Braces, their Action	33
Corrugated Plate for Floor	73
Cross Beams	63
Decoration, Constructed	83-84
Detroit Bridge Co.'s Bridge	96
Elasticity, Limit of	12
End Struts, Method of Adding to Architectural Effect	85
Eye-Bars	51
" Their Manufacture	53
Factor of Safety	11
Fairmount Bridge	83
Fastening Iron Stringers to Floor Beams	60
Fink Suspension Truss	41
Floor-Beams	63
" Factor of Safety for	16
" Riveting of	65
" Their Connection	66
Floor, General Type of, for Road Bridges	71-72
Flooring	69
" Examples of Permanent	73-74
" System	62
Floor Planks, Method of Laying	70
Floor Plates, Forms of Wrought Iron, Illustrated	73
" Wrought Iron	73
Form of Specifications in Bridge Letting	93
Framing, American System	47
Girard Avenue Bridge	84
Girders, Apparent Stiffness of	62
" Compound Riveted	34
" Depth of	64
" Solid Rolled	34
Good Bridge, Elements of	11
Gordon's Formula, Modification of	57
Gravel, Weight of	79
Guard Timbers	71
Hangers, Best Form of	65
" Factor of Safety to be Used	67
Height of Truss when Sway Bracing is Used	78
Inclined Struts, Architectural Effect	86
Invitation to Bridge Builders, Form of	93
Iron, Cast	27
" " Cold-Short	27

	PAGE
Iron, Cast, Cold-Short, Distinctive Features of	27
" " Danger from Cross Strains	28
" " Red-Short	27
" " " Distinctive Features of	27
" " Test by Short and Long Grooves	27
" " When to be Used	28
" Decarburizing of	19
" Grey	18
" Large and Small Specimens, Testing of	20
" Manufacture	17
" Ordnance	29
" Pig, Grades of	18
" White	18
" Work, Examination of	80
" " Maintenance	79
" " Method of Avoiding Rust	76
" " Painting of	80
" " Removal of Scale	80
" Wrought	21
" " Elastic Limit of	25
" " Grades of	26
Joint Box, Cast Iron, Advantages of	60
King Post Truss	35
Lattice-Truss or Double Triangular	40
Loads on Bridges	14
Loading, Table of, Proportioned to Span	15
Material, True Value of	12
Materials of Construction	17
Needle Beams	63
Panel, Length of	33
" Point	33
Pavement Blocks	74
Pig-Iron, Grades of	18
" Manufacture of	19
Pins and Eyes	49
Pins, Benders' Theory	49-50
Pin Holes, Boring of	61
Plainfield Bridge, Section of	63
" " Side View of	63
Plank for Flooring, Method of Laying	31
" Laying of	30
Planks protected against Sun-Cracking	72
Plate, Bar and Angle Iron	26
Plate Iron	26
Posts, Connection	5C

	PAGE
Posts, Sections of, Illustrated	55
" Strength of	50
" Their Resisting Power	55
Queen Post Truss	36
Rail Base, when to be Used	71
Riveted Work	45
" System, how to Use it	61
Rivets, their Pitch	65
Riveting, Hand and Power	46–47
Sap Wood	30
Screw Ends	54
Shoes or Bases	55
Sidewalks, Drainage of	71
" Method of Laying	71
Specifications for Bridges	93
Strains, Kinds of	32
Strength of Cast and Wrought Iron Columns, Table of	58
Stringer Beams	67
Stringers, Factor of Safety for	16
" Iron	68
" Tables of, Proportioned to Wheel-Loads	69
" Timber to be Used	68
" Wooden	68
Struts	33
" and Ties	33
Strut Tie	33
Testing of Bridges	88–90
Thin Webs, Precautions if Used	65
Ties	33
Timber for Stringers, Inspection of	31
" Kinds of	31
" Merchantable	29
" Preservation of	31
" Quality of	29
Timber, Season Cracks, Heart Cracks	30
Top Chord Section, Kinds of Joints	59
Top Chord Sections, { Riveted System Illustrated, For Pin Connections " }	58
Truss Bridge, its Architectural Effect	84–85
Truss Bridges	34
Trusses in Tension and Compression	34–42
Unit of Area	13
" of Strain	13
Upper Chord Section	58
Warren Truss or Single Triangular	39

	PAGE
Web, Strains in	32
" System	33
Weights of Material	78
" " Plate Iron	79
" " Timber, Table of	79
Whipple Truss, Single Canceled	38
" " Double "	39
Width of Roadway and Sidewalks	77
Wrought Iron, Characteristics of	21
" " Cold-Bend Test	24
" " Fracture of	23
" " Its Rupture	26
" " Manufacture of	20
" " Testing of	25
Zore or French Section for Floor	78

PART II.

SOLUTION OF STRAINS IN GIRDERS AND TRUSSES.

	PAGE
Action of Forces on a Beam	101
Angle and Plate Iron, Elastic Limits	118
Beams under Different Conditions of Loading	105
Bowstring Truss, Illustrative Example of Strains Solved	140–144
" ' Longitudinal Thrust	139
" " Strains in	138
Breaking Load	106
Compound Girders	113
" " Center of Gravity, how Found	113
" " Horizontal Increment in Web	115
" " " Strains in Flanges	115
" " Riveting of	115
Composition and Resolution of Forces	120
Compressive and Tensile Strains	102
Couples	103
Factor of Safety	107
Fink Suspension Truss, Solution of Example	138
" " " Strains in	137
" " " " on Suspension Rods	137
Flange Beams	109
" " Moment of Resistance of	111
Forces Represented by Lines	121

		PAGE
Formula, Practical Application of		107
King Post Truss, Strains in		122
Law of the Lever, Example		97
Lever Arm		100
Loading, Different Conditions of		109
Modulus of Rupture		104
Moment of Resistance		104
" of Rupture		104
Neutral Axis		101
Plate Girders		116
" " Allowance for Rivet Holes		114
Principle of Moments		99
Queen Post Truss, Counter Diagonal		125
" " " Example Solved		126
" " " Reactions on Abutments		127
" " " Strains in		124
Reactions on the Abutments		98
Rivets, Duty of		115
" Number to be Used		116
" Table of Sizes Proportioned to Thickness of Web Plate		118
" Value of		116
Shearing Tendency		109
Strains		121
Strength of Rectangular Beams		103
" of Stringers for Working Load		108
Table of Moment of Resistance of American Beams		112
" of Safe Center Load for Depths of Stringers		107
" of Size of Stringers for Various Spans		109
" of Web Strains Due to Movable and Fixed Loads		136
Triangle of Forces		121
Trusses, Strains in		119
Value of a Rivet Determined		117
Warren Girder, Chord Strains		133
" " Web Strains, Dead Load		134
" " " " Variable Load		135
" " Table of Strains on Diagonals		135
" " " of Web Strains		136
Web Stiffners		117
Whipple Truss		127
" " Chord Strains		128
" " " " Example Solved		129
" " Web Strains		130
" " " " Example Solved		131
Working Value of Rivets, Tables of		118
Wrought Iron Beams, Co-efficient of Safety		111

PREFACE.

It will be the effort of the writer in the following pages to point out the peculiarities of material and construction involved in the designing and building of "Iron Highway Bridges," in the hope that a dissemination of their scientific principles in a popular form, will bear fruit in a more thorough appreciation of a noble art, and in elevating the standard of requirements of this very important class of public works. The subject has been divided into two parts, each complete in itself; the one general and descriptive, and the other analytical. The former is peculiarly intended to present to public committees entrusted with the letting of bridge contracts such information as they ought to possess, while the latter is offered as an aid to engineers not experts in this branch of the profession, and yet who are often called upon to act as inspectors. The second part develops the strains in the ordinary forms of beams and trusses in an elementary manner, the principle of the lever being applied throughout, to understand which the simplest arithmetical attainments are alone necessary.

Great stress is laid upon the "strength of joints," since the essence of good bridge-building lies in their proper design. A joint must be as strong as the parts it serves to connect; as in a chain, wherein a defective link determines *its* strength, so in a bridge the absence of a necessary rivet would determine *its* strength. First-class bridge-builders recognize this relation as an axiom of their art, and it is oftentimes simply from a conscientious application of this vital principle that engineers, in making tenders for work, find themselves underbid by ignorant or unscrupulous builders, who have no other ambition than that of getting work. Ordinarily, the cheapest proposal wins the day, simply because to the average committeeman one iron bridge is as good as another, no matter from what source its plan emanates. To such a man, difference in price has no other meaning than that of being a measure of the relative greed of contractors, and he does not realize that there exist precisely the same reasons for large variations of price in iron bridges as for the difference in price between the lowest grades of shoddy and carefully woven goods. That the wisdom of such a committeeman is evidenced by a remarkable freedom from bridge accidents throughout the country is no defence for the purchase of the cheapest bridge, simply because it is a matter of exceedingly rare occurrence that a bridge is subjected to any thing near the load it ought to carry safely. The scattered travel of foot-passengers, or the uncrowded teams on the roadway do not test a bridge, and yet that is the usual condition of travel, particularly in country districts. Occasionally, circumstances arise when a bridge may become crowded, as was the case at Dixon, Ill., when, on a quiet Sunday afternoon, a Truesdell bridge fell with a horrible crash, killing and wounding many of the citizens who had congregated on that ill-fated structure to witness some unaccustomed, and therefore crowd-collecting, sight. The same story would be repeated throughout the land, were our ordinary highway bridges subjected to similar loading; and it behooves all upon whom the responsibility of buying iron bridges rests to weigh well that responsibility, and not to be deceived with the idea that their duty to their constituents requires them to erect the cheapest structure offered. There is, however, considerable difference in price for good bridges, and a good substantial bridge can be built under any of the well-recognized types of trusses. Some designs require less material than others, and the proportion of parts relating to general forms, such as depth of trusses, panel lengths, etc., still further affects the amount of material required. Two iron bridges may be built on the same general design, and they may have the same amount of metal in each, and yet one bridge is better than the other, just in proportion as the workmanship, the material, and design of the joints are better. In fact, these elements may be so poor in the second bridge as to make it positively unsafe to use, and yet to the inexperienced eye one bridge may seem almost the counterpart of the other. If this book does nothing more than bring a realizing sense of the above facts home to those public officers on whom the responsibility of carrying out public improvements usually rests, the writer will feel abundantly compensated for his labors, for he feels well aware that if this advance in official sentiment is once attained, the next step of progress will certainly follow—namely, the employment of *experts* to prepare well-defined specifications, and see that they are properly carried out.

PART I.

GENERAL AND DESCRIPTIVE.

THE essential elements of a good bridge consist in so applying the materials of construction to a given design as to have all parts of the work equally strong under the maximum loads that can ever come upon it, and that a proper relation, called the "factor of safety," should exist between the maximum loading and the strength of the structure. The term, factor of safety, as usually applied, means the number of times that the maximum load should be increased in order to break down a given structure, a ratio that varies very greatly in most American highway bridges, particularly in the "cheap ones." This conception of the term, however, is apt to be misleading, since it refers to *ultimate* strength, and not to the *limit* of *effective* strength, which last involves the idea of elasticity. The elasticity of any material is simply its recovering power from the distortion produced by the action of a force, as illustrated in the case of a rubber ball under the pressure of the hand. All materials are more or less elastic, and experiments have shown that if this elasticity is not impaired, they are not injured for use. The strain at which the recovering power of a material is destroyed is called its *limit of elasticity*, which, when once exceeded, final rupture is simply a question of time. The true measure of value, therefore, of a material is its elastic limit, and the real factor of safety is from one half to one third the values employed when the factor is referred to breaking strength, since (so far as bridge material is concerned) about that proportion exists between the force necessary to attain the elastic limit and that which produces final rupture.

When we speak of a factor of six, in the ordinary acceptation of the term, it must not be understood that a given structure can be destroyed *only* when it is loaded with six times the load for which it has been proportioned. While it may not absolutely break down until that loading is reached, its *value as a structure* is impaired the moment the material commences to be strained beyond its elastic limit, which may be the case with only double the extreme load which it has been proportioned to carry. Custom, however, has so long made use of this term, "factor of safety," with reference to *ultimate* strength, that in order to avoid confusion it will be used in that sense throughout the following pages, and if only the preceding explanation is kept in view, it makes no difference how the factor is expressed. Factors of safety usually range from four to six, the most common one being five, and it is good practice to design a bridge with two or more factors, particularly in long spans, for the reason that certain parts can only be strained fully under the extreme conditions of loading (of very rare occurrence), while others are brought under their full work almost daily, as can readily be appreciated when the subject of loads on bridges is considered.

The *unit of intensity* of a strain is expressed in pounds or tons, and the *unit of area* over which a strain acts is usually taken at one square inch, and in these units of pounds or tons per square inch, the factor of safety is applied. It has been before stated that material was uninjured when not strained beyond its elastic limit, and it might seem at first sight that the factor would be determined by dividing the ultimate strength by the elastic limit. Thus supposing an iron bar that took 60,000 lbs. per square inch to tear it apart lost its elasticity just beyond a strain of 20,000 lbs. per square inch, the apparent factor that should be used would be $\frac{60,000}{20,000} = 3$, or, in other words, the bar might be subjected to a working strain of 20,000 lbs. per square inch. This, however, would be a dangerous practice, since an allowance must be made for the imperfections of workmanship and material attending all human productions, as well as for endurance under the repeated application of moving loads. This allowance, experiment has shown, should be not less than one third greater than is expressed by the ratio of the ultimate strength to the limit of perfect elasticity. Applying this principle to the case illustrated, the factor of safety would become 4 instead of 3, and the working strain on the iron would be 15,000 lbs. per square inch, instead of 20,000 lbs.

THE LOADS to which bridges are subjected, in addition to their own weight, are of two kinds: that produced by a uniform loading extending over the whole area of the structure, and that produced by a local concentration of weight, such as may be produced by heavy stone and timber wagons, or the transport of boilers and machinery. The effect of any loading upon a bridge is further dependent on the span, for the longer the span, the greater is the fixed or dead weight, and therefore the less is the shock from passing loads felt. From this it follows that short spans should either have a higher factor of safety than long spans, or else they should be proportioned for much heavier loads. In the United States, short-span bridges are seldom built heavy enough, while, on the other hand, long-span bridges, say of 150 feet and over, are frequently made needlessly so, involving in consequence a useless expenditure.

The circumstances of location must be very carefully considered, since it is apparent that a bridge located in a country district, subject simply to the passage of occasional loads, can never be strained like a bridge in a populous community, which may be called upon to bear the incessant traffic of a city, with its processions, and often the reckless haste of a fire service. Excepting in general terms, engineers are by no means agreed as to the exact loading for which highway bridges under different circumstances should be proportioned. The usual standard is to consider a span crowded with people, which experiments have shown to vary within wide limits, depending on the density with which a given surface is packed, and the weight of the individuals with whom the experiments were made. No probable contingency, however, will pack a crowd so as to bring a heavier weight than seventy-five or eighty pounds per square foot for a general load, and for local loads it is well to bear in mind that steam road-rollers, weighing fifteen tons on an area of sixty square feet, are being introduced in many suburban towns, for the compacting of Telford pavement.* The following table, being substantially the same as was recommended by a committee of bridge experts in a report to the American Society of Civil Engineers, will be found useful in preparing specifications for road bridges, as it gives a safe and economical loading for all circumstances under which bridges are usually built:

	I.	II.	III.
	Pounds Per Square Foot.		
Span.	For city and other bridges where travel is heavy and frequent.	For towns and villages, and districts having well-ballasted roads.	Ordinary country bridges—travel infrequent and loads light.
60 feet and under.	100 lbs.	100 lbs.	75 lbs.
60 feet to 100	90	75	66
100 feet to 150	80	66	50
150 feet to 200	70	60	50
200 feet to 300	66	50	40
300 feet to 400	60	50	35

* An Aveling & Porter road-roller has fifteen tons on four wheels or rollers, each having a width of twenty inches. A roller used in England, made by the same parties, weighs thirty tons, nineteen of which are on two drivers, the width of each driver being thirty inches.

The proper *floor* strength for all spans may be obtained by considering the loads on *each pair* of wheels, for *each* roadway, and this load on bridges of the first class may be taken at from four to five tons, on bridges of the second class three to four tons, and on ordinary country bridges two to three tons. This provision for local loads may seem extreme to many, but the jar and jolt of heavy springless loads comes directly on all parts of the flooring, at successive intervals, and admonishes us that any errors made should be on the safe side.

From the above consideration of local loads on *wheels*, it follows that the cross floor-beams of a bridge are required to be of the same size and carrying capacity, whether close together or far apart, being strained alike in any case. The longitudinal stringers, on the other hand, while increasing in size for the same loads as the floor-beams are spread farther and farther apart, are independent of their distance from each other. Stringers must be of the same strength, whether spaced two or four feet apart, since any stringer may support unaided a wheel load midway between its bearings. If the wheel loads are assumed to be as high as has just been recommended, a factor of safety of four will be ample for the floor-beams and stringers, since the possibility of such loads coming upon them is very remote.

MATERIALS OF CONSTRUCTION.

In all structures affecting the daily concerns of life, to the strength of which thousands of human beings intrust their safety, the materials composing them must always be a subject of deep interest, and therefore it is of vital importance to disseminate as widely as possible a correct knowledge of their physical characteristics. And in this "Iron Age" upon which we are entering, much will be accomplished when the community realizes that in regard to *iron* at least, a "little knowledge is a dangerous thing," an aphorism applying with peculiar force to bridge-constructions. The first lesson to be learned is, that iron is a material, the qualities of which are as variable as the different localities of its production, and therefore that an iron bar is not necessarily as good if made in one place as another, simply because it is iron. Iron may be very good or very bad, or it may have all intermediate degrees of quality, and yet, to an untrained eye, a sample of the two extremes would seem to be precisely alike. It must be understood that iron is a material the most sensitive to treatment known in the constructive arts. The least, and often infinitesimal variation in the fuel, ores, and working, will result in many variations of quality, and all are more or less useful for some purpose or other. It will be the effort of the writer, in as clear and untechnical language as he can command, to point out the leading characteristics of this metal, particularly in its application to bridge purposes, and he will be abundantly satisfied if attention other than professional is awakened to the responsibility attending its selection and use.

Starting then from the ore, which is simply the pure metal combined with different degrees of earthy impurities, we have, as the first result of the contact of the ore with the fuel, the product from the blast-furnace called *pig-iron*, which commercially has different grades, numbered 1, 2, 3, 4, etc., all produced through different proportions of the fuel used, the temperature, volume, and pressure of the blast in a given time.

The low numbers are always the most expensive to produce, and are used for foundry purposes, and are known as "foundry pig," while the high numbers are converted into wrought-iron through the medium of the puddling-furnace, and are called "forge-pig." The foundry irons are often termed *grey irons*, and the forge-pig, *white iron*. Pig-iron (disregarding impurities always present) is essentially a combination of carbon and metallic iron, which combination is partly chemical and partly mechanical. Foundry pig-iron may be recognized by its softness, and, when freshly broken, by its presenting a fracture of an open, crystalline texture, and of a dull grey color. Forge-pig is hard and fine grained, generally presenting a white-appearing fracture, and at other times a mottled one. The former flows readily in the moulds of the foundry, being very fluid when melted, while the latter, which melts at a lower temperature, is somewhat pasty and flows in a sluggish stream. The operation of producing wrought-iron is simply the extraction from pig-iron of the carbon and other impurities, by means of the flame in a reverberatory-furnace, and stirring the charge of melted metal with iron bars, in order to expose every particle to the action of the oxygen of the air, which, combining with the carbon, passes off up the chimney as a gaseous product. The chemical operation thus performed is called *decarburizing*, which, were it *possible* to perfectly accomplish, and did the pig-iron contain no impurities, would result in *pure* metallic iron, which would be always alike in quality and characteristics in all parts of the world. This, however, is never the case, and there results exceedingly wide variations in the product of the puddling-furnace. Pig iron, like its namesake, who would not be driven to market, must be humored, and so metallurgists, accepting the situation, have endeavored to regulate the quality of their iron by a judicious mixture of neutralizing tendencies. In this they have been entirely successful, and all that an engineer has to do, is to say just what he wants his iron to withstand, and the service to which it is to be put, and he can have a grade of metal proper for such uses made to order. As is the quality of the pig-iron, so is that of the puddled product, which leaves the furnace as a loose, spongy-looking mass, called a "puddle-ball," still impure with cinder and slag. The next process is to consolidate the ball, and force out the impurities which are mechanically combined in the interstices of the spongy mass. This is done by hammering, or more usually by a machine called a *squeezer*, which, as its name implies, squeezes out the scoriæ, cinder and slag. The ball has now taken another shape, being consolidated into an elongated mass, of such form as to enable a still further compacting of its particles through the medium of the first set of rolls, called the roughing-rolls, to which the ball is immediately taken from the squeezer. The iron, after being passed through these rolls several times, becomes what is called a "puddled bar," and in appearance looks like a very rough and jagged-edged bar of flat iron about 20 feet long, and some $4 \times \frac{3}{4}$ inches in section. At some mills these bars are called muck-bars. They are then cut up into short lengths, and made up into "piles," according to the shaped bar it is desired to make. The piles are heated in a heating-furnace, and when at a white heat are taken out, and passed back and forth through the *finishing* rolls, from which their marketable or commercial shape is derived. This is called *best* iron, and is the degree of refinement sold by manufacturers, when simply so many tons of iron are ordered. If made from good stock—that is, well-selected pig-iron—such iron answers every requirement for ordinary purposes. But for a bridge, it is often required that this best iron should be again cut, piled, heated, and rolled into new bars, which process, while it does not change the *quality* of the iron in the least, still further refines it, and makes it more *uniform* in character, although, as may naturally be supposed, the cost of the iron is increased from ten to fifteen dollars per ton. This iron is known as " Best Best " iron. Uniformity of material is of very great importance in bridge-building—that is, if parties desire their bridges to be as strong in one part as another; and from what has preceded, it will be at once seen that this desirable end can not be obtained by open purchases in the market— that is to say, buying some bars here, and others there, wherever the different sizes can be obtained the cheapest. The temptation to such a manner of purchasing is great, in times of close competition among bridge-builders, particularly when, in nine cases out of ten, the successful bidder is such simply from being the lowest in price. We come now to speak of the distinctive physical properties of iron, and firstly of

WROUGHT-IRON.

Take a number of miscellaneous bars of best merchant iron, fracture them short off, and there will be exhibited probably as many different appearances of the fracture as there are bars. Some specimens will present coarse crystals, whitish in color, others very fine ones, of a dark gray appearance, in some lights almost black, and in others lustrous like satin. Some specimens, again, may expose a fracture wherein coarse crystals are mingled with fine. Now, what does all this express? It tells the expert that one iron is poor in quality, that it is hard, brittle, or weak, while he reads the second fracture exactly the reverse, and as that of an iron on which dependence can be placed for all purposes where strength is required. The specimen showing a combination of large and small crystals, means that the iron is not uniform in quality, and that it needed further refinement. A fractured bar tells most every thing about the quality of iron, except that of uniformity, and it exhibits this at times, as in the case above illustrated. It so happened, in the assumed exposure of fracture, that the bar was broken at a point where it lacked uniformity, but if broken a few inches either side of this point, it might not have shown any coarse crystals. Good iron that has been insufficiently refined does not show its lack of uniformity throughout the whole length of a given bar, but in spots more or less frequent, and it is simply a matter of chance if one of these *raw* spots, as they are sometimes called, occurs at the point of fracture. If, now, instead of breaking the bars off short, we slightly nick them on one side and expose them to moderate blows, so as not to bend them too rapidly, fibre will be developed in the iron of good quality, while the poor coarse crystal iron may snap off short again, after very few blows. The higher the quality of the iron, or the nearer it approaches purity, the more soft and silky will be the exposed fibre. The phenomenon of fibre can be readily understood, when it is remembered that all iron, whether pure, good, bad or indifferent, is built up, as it were, from crystals, which crystals have different degrees of fineness, depending upon impurities and the mechanical manipulations during the different stages of conversion from pig iron to the refined bar. The process of rolling develops fibre by elongating these crystals, so that a bar of rolled iron may be likened to a bundle of metallic threads of different degrees of fineness, according to the number of times the iron from which the bar has been produced has been put through the rolls. It is the ends of such threads that one observes when a bar is *suddenly* broken off short, looking as previously described, but when the bar is slowly broken, the threads, having time to arrange themselves in a new position, draw out past each other and expose fibre. It follows from what has preceded that great judgment must be exercised in criticising the quality of iron from its fracture, for crystalline fracture does not in itself indicate poor iron, nor does a fibrous one good iron. However, if care is taken to fracture the bar to be tested, under different circumstances, a fair idea can be formed of its quality and fitness for special purposes. Another method of reading the quality of iron is known as the cold-bend test, which requires no expert knowledge to understand. It consists in simply bending unnicked the bar under examination, by repeated blows from a heavy sledge-hammer, over the corner of an anvil or its equivalent, until the two sides approach each other within a distance equal to the thickness of the bar. If the iron stands this treatment without showing any signs of fracture on the back of the bend, it can be rated as of the very best quality, possessing all the requirements for bridge purposes—namely, toughness, ductility, and elasticity. This test, of course, can not show uniformity, that being a matter depending on the number of workings as before explained, and independent of quality. The cold-bend test is severer on a square bar than a round one, inasmuch as the fibres are very irregularly drawn out, being very much strained at the corners. Some very high-grade iron will even stand the cold-bend test where a screw-thread has been cut upon it, which is equivalent to numerous nickings.

The annexed cut represents the appearance of a flat and of a round bar after the cold-bend test.

FIG. 1. FIG. 2.

COLD-BEND TEST.

It was explained, under the head of the Factor of Safety, that the elasticity of a material was simply its recovering power after the removal of an extraneous force, and that so long as the limit of its recovering power was not exceeded, no injury accrued to the material. This limit of elasticity varies considerably in the different grades of iron, and generally has a value about half the ultimate strength of the iron. After the limit is exceeded, permanent set occurs, and the value of the bar is destroyed. It is probable that a certain amount of permanent set takes place in iron even under the application of very light loads, say of two or three tons per square inch, but it is so inappreciably small, being detected only by the most refined measurements, that it need not be considered in practice.

The usual method of testing a bar for its elastic limit, is to fasten to one end of the testing-machine, or to the bar itself, close to the point at which it is grappled, a rod or bar free to move at the other end, to which free end is attached an index-point. Before the strain is applied, the test-bar is scratched under this index, which mark, after the bar is put under strain, will gradually move past the stationary index, and if the strain has not exceeded the elastic limit of the bar so soon as it is removed, the mark will return to its former position under the index. Successive applications and removals of the strain are required usually to determine the elastic limit, else it might be unwittingly passed under a continually increasing power. After becoming satisfied as to the elastic qualities of the bar, a final application of the strain can be made in order to tear the bar in two, care being taken to note how much it stretches before final rupture. This process of stretching to rupture, exhibits not only the ductility of the iron, but also the degree of uniformity, shown by a greater or less inequality in the amount of stretching at different portions of the bar.

The beauty of the cold-bend test is, that it shows simply and inexpensively the same qualities (excepting uniformity) that the testing apparatus measures in pounds and inches, and for practical purposes nothing else is needed. The result of many thousands of experiments on American irons shows that for bridge purposes, bar-iron should stand at least 50,000 lbs. per square inch before rupture, should have an elastic limit not less than 20,000 lbs. per square inch, and should elongate at least twelve per cent of its length (or 1½ inches to the foot), before ultimate strength is reached.* Most of the first-class bridge-builders use a higher grade iron than the above, which is given simply as a *minimum* quality for highway-bridges, easily attainable. Angle-iron and plate-iron, as *usually* applied, are from ten to fifteen per cent weaker than good bars, and, therefore, bridges built from such irons should have proportionately just so much excess of metal over bridges built from bars, a requirement that the buyers of iron bridges, in this country at least, have not as yet learned to insist upon. Before passing from this subject, it should be remarked that the tables of strength of wrought-iron are based upon experiments made on small bars, having cross-sectional areas of about one inch. Large bars will not show the same ultimate strength that small ones do, of the same make, a fact that must be borne in mind when specifications are being prepared. For example, the *same iron* in a bar having one inch area may require a strain equivalent to 10,000 lbs. per square inch to rupture it, in *excess* of that required when formed in a bar having an area of four or five inches. Until a comparatively recent date, no attention was paid to the effect of the form of the specimen to be tested. Test specimens are simply short pieces of iron, three or more inches long, the middle of which is grooved down to exact gauge, and which be-

* Very accurate gauging under a magnifying instrument will indicate a permanent set long before 20,000 lbs. per square inch is reached, and probably

comes the area to which the breaking strain is referred. The character of the grooving, whether long or short, affects, in a marked degree, the result of a test. If the groove is a short one, the iron will break at a much higher strain per square inch than if it had been long, and this result is due to the fact that a free stretching of the fibres is prevented by the reinforcement derived from the metal contiguous to the ruptured section of the short-grooved specimen. This difference, due solely to the preparation of the specimen, will amount in some cases to as much as fifty per cent. The explanation of this apparent anomaly in the strength of iron may be made still clearer by an inspection of the cut, where Fig. 1 represents a long-grooved specimen, and Fig. 2 a short-grooved one. The shoulders at either end are formed for the grappling-irons of the testing-machine.

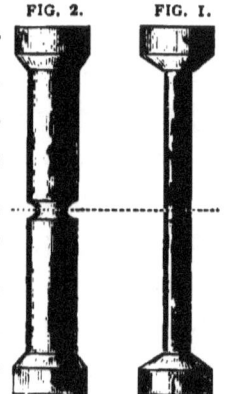

FIG. 2. FIG. 1.

SHORT AND LONG GROOVED SPECIMENS.

There are two terms continually met with among iron-workers—namely, *red-short* and *cold-short* iron, which it may be advisable to explain. The peculiarity of the former is, that while very strong and tough when cold, it is difficult to work in the forge except under very high heats, otherwise it will crumble and waste, and for this cause has received the enmity of smiths. On the other hand, cold-short iron is brittle when cold, and absolutely unsafe to use where life depends upon its

with not over a ton strain—but ordinary methods, such as are used in the shops, will not detect a set below 20,000 to 25,000 lbs. per square inch.

integrity. The smith likes to use it, since it works and welds readily in the forge at low heats. The best manufacturers aim to have a neutral product, which, if it has any tendency at all, is on the side of red shortness.

Cast-iron in bridge-building is so little used at the present day, except in the form of bearing-blocks, post-caps and bases, or washers, that little need be said about it. In its very nature, it is a brittle material, and even while apparently doing good service, may be dangerously near failure. It has an irregular elasticity, and in cold climates it has been known to fracture through the freezing of water that had found its way into unprotected cavities. In the form of *long columns*, it is of course very inferior to wrought-iron. Such columns are exposed to cross strains, and have a tendency to fail by bending and not by crushing. Tension in some part always accompanies a cross strain, to resist which cast-iron is a very uncertain material. Castings may have initial strains through unequal cooling, or they may be thinner on one side than another, or they may be weak through concealed holes, "cold shuts," or cinder. No human foresight can remove these risks; and especially in bridge-building is it important to reduce all risks to a minimum, and for this reason, if for no other, cast-iron should be discarded for such purposes, except in those places where it would be very expensive to forge wrought-iron, places where none other than a direct crushing strain can ever occur, as previously instanced. The iron from which castings are made should be selected with great care, and it should have sufficient meltings, 2 to 4, before being put into its final shape. Such castings, when broken, should present a fine-grained grayish fracture, and their skin should be generally smooth, but not smooth like stove-plate castings, as such iron is very unsuitable where strength is desired. Stove-plate castings must be made from a very fluid iron, one that runs thin, and sharply fills the moulds, and such irons are very weak. Ordnance iron, with a *tensile* strength occasionally equal to that of inferior wrought-iron, is the best cast-iron possible to have, but it is expensive, and rarely used on that account. Such a grade of iron, however, should always be insisted upon where bridges are permitted to be built having cast-iron top chords and posts.

TIMBER.

Whatever modesty is shown through conscious ignorance in criticising iron and its fabrication, it quickly disappears when the question of timber is under consideration, almost every one being positive as to what is good timber, and very frequently unreasonable exactions are imposed. The main trouble that arises, in the execution of contracts, arises from the interpretation given to the term *merchantable*, an expression somewhat vague, without other limitations. All bridge-timber should be *sound*—that is, free from loose or black knots, heart-cracks, and wind-shakes, and it should not be cut from logs obtained from dead trees. Seasoned timber, especially when it has been exposed to the direct rays of the sun during the process of seasoning, is apt to have more or less cracks, called season-cracks, which must not be confounded with heart-cracks and shakes. They can be distinguished from each other from the fact that the cracks due to seasoning are sharp, while those due to shakes are splintery—the splinters, in many cases, being easily torn off. Well-seasoned timber wears much longer than green timber; but since bridge-plank is seldom, if ever, kept in stock, and since public works rarely have their needs anticipated, lumber is almost always procured fresh from the mills. The durability of timber would be very much enhanced if kept soaking in water for a few months after it is cut into plank, after which seasoning proceeds very rapidly, the water having acted as a solvent in ridding the pores, to a great extent, of sap and nitrogenous matter, the decaying elements of wood. Sap-wood—that is, the wood newest made and next the bark—is not desirable, as it will wear away faster and decay sooner than the heart-wood, but practically it is impossible to obtain timber of any size and in large quantities entirely free from it, unless at a very great increase of cost. Sap-wood may be recognized as being lighter in color, softer, and of more open fibre than the heart-wood. Timber is regarded as merchantable when it has not more than three sappy corners, although some inspectors do not permit of more than two; but as bridge-plank usually wear out before they rot out, a latitude can with propriety be observed here, and the plank laid with the sap corners down, thus: , the dark portion representing the sap-wood. Wane or bark edges are very apt to occur in otherwise first-class sound timber, but should not insure condemnation if only on *one* corner, if the plank can be laid with that corner down. If on two under corners, the plank would be next to the slab (or outside cut), and therefore almost all sap-wood, and should not be permitted to pass by the inspector. For stringer timbers, inspection ought to be somewhat more rigid than for floor-plank, but guided by the same common-sense principles, and the farther consideration, how much surplus strength the stringers possess. The kinds of lumber used are mostly oak and pine, both white and yellow; to these may be added, for plank purposes, beech, birch, and maple, and occasionally spruce, when two courses of plank are used, the upper one being of hard wood. All things being considered, the writer prefers close-grained yellow pine for floor-planks, it being much less expensive than a proper quality of oak, and besides less slippery for horses in frosty weather. As to artificial means for preserving timber, a number of processes have been tried with success. The various methods of creosoting and burnetizing are the more common in use. The city of Boston required the latter process to be applied to spruce plank in some bridges recently built, as one eminently effective and cheap. Any process used, unless thoroughly well done—that is, unless all the pores and cells are *filled* with the preservative material—is even detrimental, since in such cases dry rot inevitably sets in at an early day.

KINDS OF BRIDGES.

The various kinds of bridges ordinarily met with may be classed under one of four heads, namely, the plain beam or girder, the beam truss, the suspension truss, and the arch truss or bowstring. The first class needs no explanation. The second form includes all trusses where both top and bottom chords are absolutely essential, while the third embraces those trusses wherein only the upper chord is essential. The bowstring is properly not a truss at all, but simply an arch wherein the horizontal tie takes the place of fixed abutments. The office of all girders, whether plain or trussed, is to transmit weight to the points of support, which action develops two classes of strains, namely, horizontal and vertical (sometimes called shearing). The former are resisted by the top and bottom longitudinal chords or flanges, while the latter are taken up by the intermediate bracing, called collectively the *web*, which applies to all the material lying between the chords or flanges, whether open as in a truss, or solid as in a plate-girder. The longitudinal strains in the chords are either *compressive* or *tensile*, and whichever may be the case, the *quality* of the strain is the same throughout the chord considered. The web is exposed to both kinds of strain, the parts of which, if a truss, are alternately in tension and compression in the march of a given weight to either abutment. The tension members of the web are called *ties*, and they may be either vertical or inclined. The compressive portions of the web are called *struts*, or *posts*, and may also be vertical or inclined. When ties are vertical, the posts are inclined, and *vice versa*, or both may be inclined. *Strut tie*, as the name implies, means that a web member may act either by tension or compression. The point where a tie and a strut intersect in a chord, is called a panel-point, and the distance between two such points is called a panel-length. Again, a portion of the web system are called *main* braces, or ties, and a portion *counter* braces, or ties. The former embrace all parts of the web which carry that part of the weight going to the nearer abutment either side of the centre of the truss, and are lightest toward the middle and heaviest toward the ends of the span, while the latter run in a contrary direction to that of the main braces, and carry that portion of the load going to the farther abutment, and they are heaviest at the centre and least at the ends of the span. Main braces may be made to act as counters, if they are constructed to act either by tension or compression. The office of the "counters" is simply to prevent distortion or change of form in a truss, and they are only necessary when the truss is subjected to the action of a variable load, as is the case on all bridges. They can only act when the main braces to which they are opposed are relaxed, and then have an action equal to the difference between the effects of the variable and fixed loads, acting in opposite directions. In a bridge very heavy in proportion to the moving load, this excess is soon lost either side of the truss centre, when the counters can of course be left out. Ordinarily they are continued a short distance beyond theoretical requirements, in order to diminish vibration, which they materially assist in preventing when screwed up tightly. The usual forms of truss bridges are illustrated by the succeeding figures, on each of which is represented, by means of lines of varying width, not only the parts strained the *greatest*, but also the *kind* of strain. Tension is shown in fine lines, and compression in full black ones. The weights producing strain are supposed to be located immediately at the panel-points, the whole materially aiding the mind in forming a very fair idea of how trusses really do act, when coupled with the descriptions and definitions just given.

Figs. 3 and 4 represent plain girders for short spans, in which the flange and web parts are noted. Such gir-

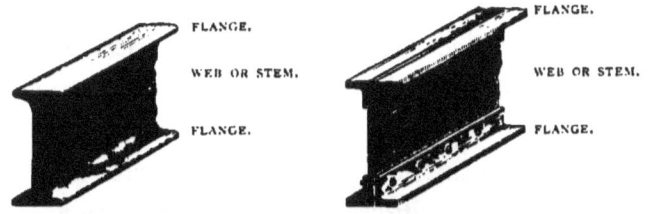

FIG. 3. SOLID ROLLED BEAM.   FIG. 4. COMPOUND RIVETED GIRDER.

ders are often used of solid section, and are called rolled beams, being finished ready for use direct from the rolling-mills. They are made of varying sizes and weights, from the four-inch beam, weighing 30 lbs. per yard, to the fifteen-inch beam, weighing 200 lbs. per yard. These beams, however, are more expensive than the compound riveted girder, made with plates and angle irons, but are 10 per cent stronger. The riveted girder can be made of any depth, and is therefore adapted for much longer spans than the rolled beams.

Fig. 5 shows the simplest form of truss, and consists of a post and two inclined ties supporting the middle of a beam, that would otherwise be too weak to sustain a load. This supporting system in effect halves the span, the post performing the office of a pier, carrying one half the load of both subdivisions of the beam.

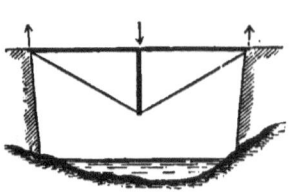

FIG. 5. KING POST TRUSS.

Now, since all the load must finally rest on the two end supports or abutments, that portion that rests on the post can only reach them through the medium of the inclined ties, intersecting at its foot, each tie taking up half the load carried by the post. These ties are strained in excess of the load they transmit to the abutments in proportion to their deviation from a vertical line ; or, in other words, an inclined pull requires greater effort than a direct one, as almost every one has experienced. Whenever a force is exerted at an angle, a *horizontal effect* is *always* produced, and in proportion to the angle at which it is applied. The flatter the angle, the greater the horizontal effect, and *vice versa*. In the truss before us, the abutment ends of the inclined ties, by virtue of this horizontal effect, pull toward each other, producing compression in the horizontal beam to which they are attached. This form of truss is called the "King Post" truss, and when inverted will be at once recognized as the commonest form of wooden trussing in existence. In that case, however, the vertical post becomes a tie, the inclined ties become thrust braces, and the beam is strained *tensively*, instead of compressively, since the horizontal effect of the inclined thrust braces is to tear the beam apart.

Fig. 6. When an opening becomes too great to be spanned by a beam trussed with a single post, two posts

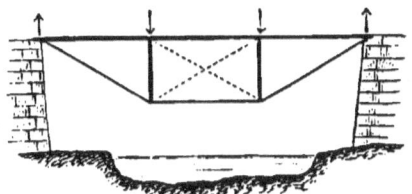

FIG. 6. QUEEN POST TRUSS.

are added, forming three spans, the posts being the piers as before, which piers are supported in turn by the inclined ties running up to the ends of the horizontal beam as before; each tie sustaining the whole weight on one pier or post. This is a complete truss when both posts are loaded; but if only *one* is loaded, the condition of affairs changes. The load is unbalanced on the other side of the centre, and the horizontal effect of the inclined tie on the loaded side will be greater on the beam (which hereafter we will call the upper chord) than that from the similar tie on the unloaded side. The result will be a distortion of the frame, the loaded post sinking and the unloaded one rising. All that is necessary to prevent this destructive effect, is to enable that portion of the load that *must* be carried by the further abutment, to go there by the most direct route, which is manifestly through the medium of a diagonal tie from the *foot* of the loaded post to the *top* of the unloaded one, or a diagonal strut from the *top* of the loaded post to the *foot* of the unloaded one. This diagonal is the counter-diagonal previously defined. Its introduction in the elementary truss just described, and known as the " Queen Post Truss," is a pointed illustration of the value of the triangle in trussing, which is the only geometrical figure that resists change of form. Inverting this truss, as was done before with the King Post, we have the Queen Post in a more familiar shape; and while the effect of the loads on the several parts is precisely the same in *amount* of strain engendered, the quality is reversed. That is, the upper chord becomes now the lower chord, and suffers tension, the inclined ties become thrust braces, the posts change to ties, and the lower chord becomes now the top chord undergoing compression. The counter-diagonals also become reversed as to tension or compression.

The forms of trusses just described embrace all the elements of simple trussing, and an extension of these principles is all that is necessary to meet the ordinary requirements of every-day practice. By adding to the number of posts, the Whipple or Pratt truss, Figs. 7 and 8,

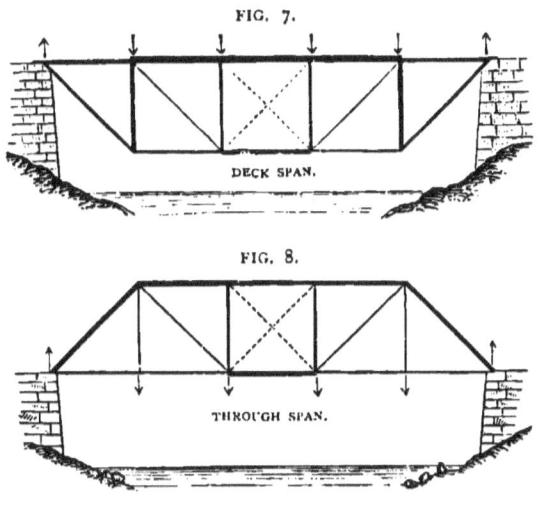

WHIPPLE SINGLE CANCELLED TRUSSES.

is formed, a plan of truss the popularity of which is well deserved. The inclination of the end posts, though not essential, results in a saving of material over vertical posts, but the latter form produces, in the judgment of many, a more pleasing effect. A study of the diagram will show how cumulative the horizontal effects of each diagonal main tie are toward the centre of the chords; and also how it is that any main tie must carry *all* the weight between its own panel load and the centre of the span.

Fig. 9. When a span becomes very long, and it is constructively and economically inconvenient to have one system of triangles, two systems are introduced, complete and independent of each other, each one being

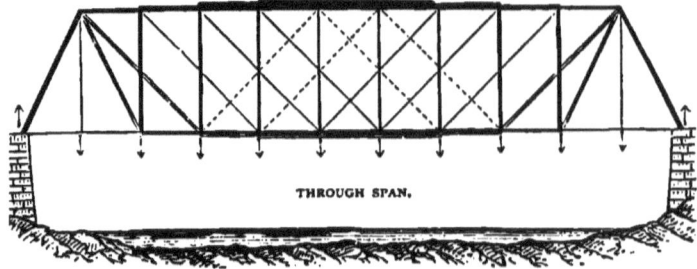

FIG. 9. DOUBLE CANCELLED WHIPPLE TRUSS.

formed of triangles having bases of two panel-lengths. The principle of the Queen Post still holds good as before, as it would do if there were three or more series of triangles, each series doing its own work in transmitting the loads to the abutments, independent of any other. The truss illustrated in Fig. 9 is known as the double cancelled Whipple or Quadrangular truss, and has been used in spans of over 400 feet.

Fig. 10 appears, at first sight, a greater variation from the elementary truss form previously described

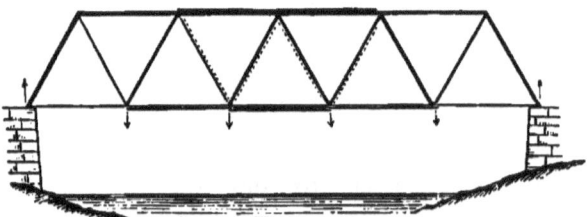

FIG. 10. SINGLE TRIANGULAR OR WARREN TRUSS.

than is really the case. The principle of the triangle being here developed to its utmost perfection, this form is usually known as the "Triangular" truss, although sometimes called the "Warren Girder." The marked difference between this form of truss and the Whipple and Queen Post trusses consists in the fact that the posts as well as the tension-rods are inclined, and if the angle of inclination is well proportioned, a considerable economy of material is obtained over that required by the straight post trusses. When a *vertical* post is used, the weight delivered to it by its tension-rod makes no progress whatever toward the abutment; but in the case of an inclined post, by the time the weight has been transmitted to its foot, it has progressed toward the abutment by an amount equal to the horizontal reach of the post. When the span becomes long and the stretch of the triangles is so great as to necessitate an intermediate support for the flooring, a rod is dropped from the apexes of the triangles to form such support,

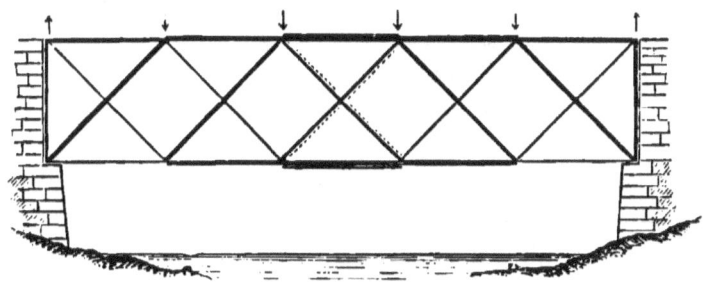

FIG. 11. DOUBLE TRIANGULAR OR LATTICE TRUSS.

or two systems of triangles may be used corresponding to the double cancelled Whipple truss, as in Fig. 11. In the case of the trusses being *beneath* the roadway, the vertical rod becomes a post, as the load then presses from above, instead of being suspended from below. In this form of truss, it will be noticed that the horizontal effect at each panel-point is made up of two portions—one due to the thrust of the posts, and the other to the pull of the ties, both being inclined and acting in the same direction—just as a man pushing behind a wagon adds to the effect of a man pulling it in front. While the vertical post truss has only one increment at each panel-point, yet for the same depth of truss the sum of all the increments on either system will be the same at the centre of the chords.

Fig. 12 illustrates the suspension truss, where only a top chord is essential, and is nothing more than an

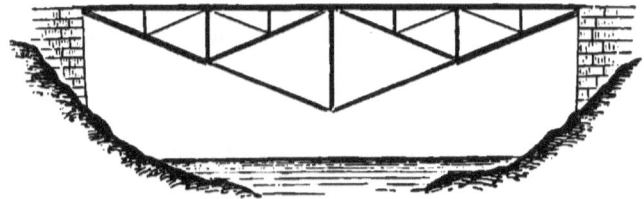

FIG. 12. FINK SUSPENSION TRUSS.

ordinary roof truss turned upside down. This form was first developed for bridge purposes by Mr. Albert Fink, and it almost universally goes by his name. It is developed from the elementary truss, Fig. 5, as will be apparent on inspection. By imagining the King Post truss in Fig. 5 to become so long as to require intermediate support, it is accomplished in this case by adding sub-systems, acting precisely like the main system, only in a minor degree. The load on each post splits in half, as it were, at the post-foot, each portion being carried up the inclined ties to the top of the adjoining posts, each minor system thus *adding* to the weight imposed on the next larger system, until the whole load is finally delivered to the abutment. The main system extending over the whole span is called the primary system; the systems extending over each half span are secondary systems; those over each quarter of the span, are tertiary systems; those over each eighth of the span, quaternary systems, and so on. The horizontal increments of all the ties accumulate at the extreme ends of the top chord, producing uniform compression throughout its whole length.

Fig. 13 is the familiar bowstring, which acts, as before remarked, like an arch, and bears no relation what-

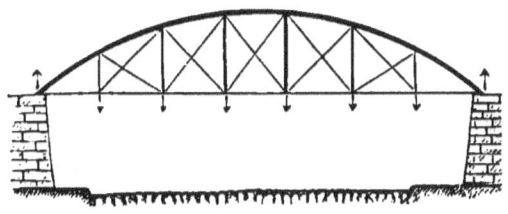

FIG. 13. BOWSTRING TRUSS.

ever to the typical form of trusses developed from Figs. 5 to 12. The essential parts are the bow and tie, the latter taking the place of fixed thrust abutments. The web for a uniform load need be nothing more than vertical rods, carrying simply the separate loads at the panel-points. Where the load is variable, as is always the case in bridges, and if the arch is not stiff enough in itself to resist distortion, diagonals must be introduced in the web performing simply the office of counter-braces. Like them, they are strained the greatest in the centre of the span and least at the ends.

THE SELECTION OF BRIDGES

should be governed by economy and adaptability to location, since no one of the well-recognized types of bridges is better than another. Apart from such motives, *any* bridge designed on correct principles is a good one, whether a beam-truss, a suspension-truss, or a bowstring. On the contrary, any one is bad if improperly designed, and the principles of its construction ignorantly conceived. A general rule that will lead to satisfactory results is *to ignore any plan of bridge that can not be accurately analyzed as to the character and amount of strain occurring in all its parts*—such, for instance, as the Truesdell bridge, scores of which have been built during the last fifteen years; and assuming that the great majority are still in use, giving satisfaction to their users, yet their form of construction is one that removes them beyond the pale of the most refined analysis. They are purely empirical structures, and being such their construction should under no circumstances be permitted. It is bordering on criminality to build any structure on a plan that no human being can tell definitely any thing about, when there are so many plans that we thoroughly understand.

METHODS OF CONSTRUCTION AND FORMS OF SECTIONS.

The various systems under which iron-work is framed may be classified as the "pin connection," "screw-end connections," and all "riveted connections," which may be and often are combined, to a greater or less extent, in the same bridge. The first two systems are peculiarly American in their origin and practice, while the last is the system pursued almost entirely in England and on the Continent, although latterly the attention of American engineers has been drawn to a considerable extent to riveted work. As has been before intimated, the knowledge of a good bridge-designer will be shown in his details, more than in his mathematical expertness in figuring up strains; and, perhaps, it will not be hazarding too much to say, by way of emphasizing this remark, that few iron highway bridges built in the United States are as strong at the joints as the parts they serve to connect. The very great difficulty in obtaining this joint strength in purely riveted work is due to the general nature of such designs. In the first place, as built in this country, the bars or pieces uniting at the panel-points *do not assemble in the axial lines* of the truss, thus producing a complexity of cross strains unknowable in amount. In a large bridge, involving heavy pieces and large joints, it is *impossible* to so dispose the rivets as to distribute the strain equally among them, although they can only be proportioned on that supposition. It will be apparent to any one, on a moment's reflection, that when two pieces of iron are riveted together through the medium of a splice-plate, the rivets at the ends of the splice are the *first* ones to feel the effect of a strain in the bars, and consequently are brought into action before the rivets at the middle of the splice are affected ; and if the bars are large, the splice-plate long, and the rivets numerous, it is doubtful if the rivets in the middle of a splice do any service whatever; certainly not before the iron has stretched considerably, in which case the first rivets may have upon them double the strain they were calculated to bear. As manufactured in this country, the holes of each piece are separately punched from wooden templates, and despite all the care exercised, the drift-pin must be always at hand to *force* the matching of the holes of contiguous plates, to admit the insertion of the rivet, thus developing initial strains on the iron impossible to compute, which may be regarded as another very serious indictment of riveted work. Workmen can not always be watched, and the eyes of even the fiercest inspector can not keep every hole and rivet before him. The carelessness of a workman may be rapidly and nicely covered up with a neatly-shaped rivet-head, which tells no tale of the horribly mutilated holes beneath, to which a cold-chisel had possibly been applied, or perchance the holes overlapped too badly for the drift-pin to even give an appearance of matching. Another imperfection very apt to creep in when hand-riveting is employed, and one, too, that is so thoroughly concealed as to be impossible of detection, is the *imperfect filling of the holes*. The chances of such a serious defect increase with the number of the plates riveted together, and owing to the shrinkage of the hot driven rivet-heads, they bind so closely to the surfaces of the outer plates, that striking with a hammer to test "looseness" is a very fallacious test. The high strain under which rivet-heads are left through shrinkage in cooling is often shown by their apparent brittleness when cut off by a cold-chisel. They will at times snap off like a piece of glass under the first blow. A hand-driven rivet will very frequently drop out from its own weight, when once the head is knocked off, showing that the shank of the rivet shrinks away from the holes, and when this is not the case, they are as apt to retain their position through the distortion caused by unmatched plates as to a perfect filling of the holes. In Europe, where the riveted system has been developed to its utmost perfection, these inherent defects are recognized, as is shown by the great care with which their riveted work is manufactured, such as *drilling* the rivet-holes through the plates and pieces to be joined while clamped in position, and thus overcoming almost entirely the evil effects of drifting and distorted rivets. Power-riveting is largely employed, as by that mode alone there is any reasonable certainty of filled holes. Did American girder-shops pursue the European system, our riveted bridges would cost much more than they now do, and they would be proportionately better. To do this, however, requires more than the customary standard plant—namely, a punch, a pair of shears, and drift-pins, which any old boiler-shop can furnish. Before leaving the subject of riveted work, it is well to call attention to "field-riveting"—that is, where spans are so large that they must be shipped in parts, which are riveted together in the final position of the work. Whatever objection has been urged against *shop*-riveting is intensified in a high degree when the field-riveter steps in to do his part of the work. He must work in constrained positions and in all sorts of weather. If the work in the shop has been well done, that in the field is pretty sure to be badly done; and as this last applies principally to the *joints*, the most vital parts of the whole structure, the work must be judged entirely by them. In contrast with riveted work, we have the *machine*-made bearings and connections, which may be attained either by means of *pins* or *screw-ends*, or a combination of both. It is through the adoption of this constructive idea that the Americans have been able to surpass the rest of the world in bridge-building.

This American system, as it is universally called, permits of the most economical use of material possible, is wonderfully well adapted for long spans, and enables the engineer to select the quality and shape of material best adapted for any given portion of his design. It is a system that permits of closer harmony between theory and practice than is possible to attain in the European method or its American imitation, concerning which enough has been said to show how lamentably deficient that system is in this particular. In a bridge on the American system, the strains, being axial, coincide with the skeleton diagram of the truss, and, further, the strains can be accurately computed, and need have no more material provided to meet their action than is absolutely necessary. The more usual mode of connection in this system is by means of *pins*, which joints, when well designed and executed, leave nothing to be desired. The main points to be considered are the sizes of the pins, the reinforcing of the upper chord and post-bearings, the fit between the pins and eyes, the proportion of the heads of the tension-bars, and the uniformity in lengths of similar parts in each panel. It is no part of a book of this character to give specific rules for the proper proportion of these parts, but the great importance of the subject, and the fact that the majority of American highway bridges are very deficient in "joint proportion," warrant an attempt to make clear the requirements of the pin-connection. Pins can not be made too large, and are governed in size by the largest tension eye-bars through which they pass. These occur in the lower chord or in the main diagonals at the ends of a truss. Whether a pin is a half inch more or less in diameter is an economy not worth consideration—only be sure that the error, if any, is toward the larger diameter. Considering the very great importance of properly proportioned pins, it is somewhat remarkable that so little attention has been given to the subject. For years the crude conclusions of Sir Charles Fox, drawn from very meagre experiments, made more than a dozen years since in England, have been a sort of blind guide for engineers. They have been supplemented within the last five years by the experiments of Mr. Berkeley, also English; and although these last and more complete experiments have shown how erroneous Sir Charles Fox's rules are, yet those rules are still given in modern text-books as proper practice. Mr. Charles Bender, C.E., has very ably investigated the subject theoretically, and shows the various influences operating to modify the size of pins, according to the position of the different bars assembled upon them, and he shows the fallacy of deriving rules from experiments made upon bars having a uniform ratio of width to thickness, or on pins only exposed to direct shearing action. The best experiments and theoretical investigations go to show that the size of pins for flat bars should be not less in diameter than $\frac{8}{10}$ the width of the bar, and for square bars their diameter should be not less than $1\frac{3}{4}$ times the side of the square. It will be noticed that these proportions result in pins enormously in excess of what would be necessary for simple shearing. For example, a bar 4 x 1 requires a pin $3\frac{1}{4}$ inches in diameter, the area of section of which is $8\frac{1}{4}$ square inches, while that of the bar is but 4 inches. Pins should be carefully turned to gauge, and fit the holes through which they pass with the least play with which it is possible to put the work together, which the best practice has established at about $\frac{1}{64}$ of an inch. Of as much importance as the pins, is the proper form to be given to the ends of the "links" or "eye-bars," the name usually given to the braces and lower chord bars. In order that the pin will not tear through the eyes before the body of the bar is at the point of rupture, experiment has shown that the link-heads must be full, and of gradual curvature, the proportions of which being dependent somewhat on the mode of manufacture. Still further experiments are required on eye-bars of various sizes, to determine with accuracy just what proportion should be given to the heads; but so far as experience has gone, it points to a proportion in the case of flat bars of about 50 per cent of metal *through* the pin in excess of that through the body of the bar, and in front of the pin about the same as is contained in the body of the bar. Back of the pin, the curve uniting the head with the body of the bar should be a gradual one, so that the strain in the bar will not be too abruptly transferred around the pin. The annexed cut represents the end of an eye-bar, with a pin passing through it, the relative intensity of the surface pressure being indicated in shaded lines. It will be perceived how important it is to have tight-fitting pins, since the first pressure is simply a line of contact, the semi-circumference of the pin only coming into bearing when the pressure has upset the metal in front of the pin by an amount equal to the extent of play in the eye. Some engineers consider that the bearing surface should be determined by projecting the semi-circumference on the diameter, allowing nothing for frictional resistance when the pin and eye surfaces are in full contact. In that view of the case for flat bars, with heads uniform in thickness with the body, pins should have a diameter equal to the full depth of the bar; or, in case it is unadvisable to have such large pins, the required bearing area can be made up by thickening the eyes.

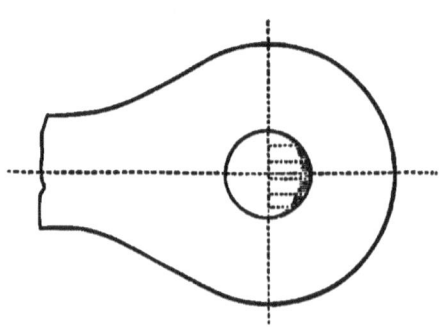

FIG. 14. LINK OR EYE-BAR HEAD, SHOWING RELATIVE INTENSITIES OF PRESSURE ON PIN.

When square bars are used, the eyes should be formed by long loop-welds, which gives, of course, ample material around the pin, being equal to the side of the square. Round bars should be forged with an equivalent flat head, as it is impossible to properly loop-weld a "round," and have a satisfactory flat bearing on the pin. The eyes of all links should be carefully bored to match the pin, with minimum clearance compatible with erection of the work. Since in all link bridges each individual bar is calculated to perform a given proportion of duty, uniformity of length, particularly in bars of the *same panel*, is of the first importance. Otherwise, an inequality of strain will result after the work is erected, the tighter bars taking all the load at first, only bringing the slack ones into play after they have stretched a sufficient amount so to do. These errors of length creep in from two causes—

namely, carelessness in centring the eyes from the master-gauge, and the variations of temperature at which they are bored. The best shops use double-end boring-machines mounted on wrought-iron beds, and if care is taken that the bars have been long enough lying in the same temperature as that of the machine, the second class of errors are removed to a remote possibility; the avoidance of the first being simply a matter of shop system, in checking measurements and using intelligent supervision.

Flat eye-bars (the form now almost universally used by the best designers) are manufactured in America, either by welding the eyes previously forged into shape to the ends of the bars, by die-forging under a steam-hammer, or upsetting by means of steam or hydraulic power. The former process is purely a welding process, and should be performed with great care in a hollow coke fire, the form of weld known as the split weld being used. The second process is a weld to the extent that a slab is forged down on the ends of the bar under the powerful blows of a steam-hammer, the shaping being performed at the same instant, the anvil and the hammer having matched die-faces, while the latter process consists in forcing the ends of the bar itself into properly-shaped moulds or dies under an intermittent or a steady, continuous pressure of a ram, the ends being previously heated to a white heat. All these processes are in use, and have given satisfaction; but the two latter have decidedly the preference among engineers, owing to their greater reliability. The second method is the most flexible, in that there are no such limitations of ratio of width to thickness as the direct upsetting process necessitates. One fact in regard to upset-bars must not be overlooked, and that is the distortion of the fibre, and consequent change in the character of the iron. This is sure to be extreme, if the operation is performed under too low pressure, or if the bar is heavy and the head large, in proportion to what may be called the mass of the upsetting-machine. Where bars are wide and thin, there is a very great distortion of fibre since a large amount of iron must be forced back to fill the moulds, and which a slight etching with acid will develop very clearly. It is owing to the deterioration of the iron in the heads of upset-bars that American experiments have resulted in somewhat different proportions from those made in England on bars of English manufacture. Whether upsetting is done by repeated impact, or by the steady, continuous pressure of the hydraulic ram fed from an accumulator, there is a marked difference in the result. Iron is most susceptible to change of form without deterioration when operated upon in a highly heated state, and since a bar commences to cool the moment it is taken out of the furnace, the most rapid means of shaping it will injure it the least. A fibrous bar, operated upon in a cold state, will be so modified in its molecular arrangement as to 'become crystalline. Again, in operating upon the end of a bar, just from the heating furnace, it must of course be firmly gripped behind the die, and where the iron is comparatively cold. In the case of slow upsetting by impact, the iron is gradually crowded back from the soft end, the effect of each blow being less and less as the metal gets cooler and the fibres become compacted. At the end of the operation, the metal will have chilled off rapidly, and near the base of the upset be almost cold. At the point of "grip" the metal becomes more or less crystallized, according to the temperature at that point. In view of this effect of temperature on iron, it follows that upsetting should be only performed by continuous pressure, by means of which the iron may be driven back in the die at welding heat, at one stroke of the piston.

Screw-ends are sometimes used for the upper ends of the diagonals, and form their connection with the top chord through the medium of a casting, which requires a very awkward and ugly enlargement to admit of their passage. Screw-ends should be enlarged over the body of the bar by upsetting, so that the cutting of the screw-threads will not diminish the sectional area. A serious objection to the use of screw-ends arises from the fact that they are a temptation to those custodians of public works who have a mania for screwing up any thing they can get a wrench around, and so, in their efforts to "adjust" a bridge, they are very apt to leave the diagonals under different degrees of tension. To adjust screw-ends properly, the workman must combine the "feel" of the wrench with the striking of the bars, so as to judge of the tension by the sound, which involves somewhat of a musical ear, not possessed by every mechanic. Practically it is impossible to tap nuts so that they correspond with the threads of the screw. The dies will wear, no matter how carefully they may have been hardened, and the hardening process itself must affect the character of the threads.

The *post connection* with a pin is made through the medium of " shoes " or " bases," either of wrought iron or cast, or both combined, depending on the form of post used. The bearing on the pin, or, in other words, the thickness in inches of that portion of the shoe through which the pin passes, should be not less than the compressive strain (as exhibited in the line diagram of strains which ought to accompany all proposals) in pounds, divided by twelve thousand times the diameter of the pin. The sections of posts in ordinary use are exhibited herewith in the order of their relative theoretical merit. The

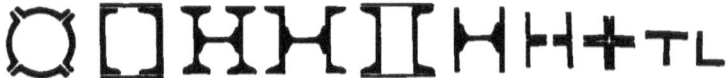

FIG. 15. SECTIONS OF POSTS OR STRUTS.

first is the Phœnix hollow column; the next four are made from solid rolled sections, and the last and weakest are compounded sections, as shown.* The resisting power of posts is based upon the ratio of their length divided by their diameter, and also upon the fact of their having *round* or *square* end connections. For the first five sec-

* The last four sections are the forms of struts used in riveted work, and it needs not the eye of an expert to realize that they are immeasurably inferior to any of the preceding sections.

tions, the diameter to be taken, in determining above ratio, is the *least* side of the *least* rectangle with which they can be circumscribed. For the other sections, two thirds of the least side must be taken for the diameter. While this method of determining effective diameters is not absolutely accurate, it is sufficiently near the truth to test the merit of competitive designs. Where a post bears directly on a pin, it should be regarded as having a round end. There is probably no property of iron about which less is positively known than its real strength when in the form of posts or columns. Certain general laws have been determined by the experiments thus far made, among the most important of which are the following:

The strength of a column with square end bearings being called unity, that of a column with *both* ends rounded (like the ends of an egg) will be one third, and that of a column with one end square and one end round will be a mean between the first two. That is, the numbers 1, $\frac{2}{3}$ and $\frac{1}{3}$ represent the relative strength of columns, according as the bearings are square, one round and one square, or both round. The formula mostly in use for computing the strength of posts is an empirical one, invented by Lewis Gordon, of England, and is based upon the experiments made for the British Board of Trade, by Eaton Hodgkinson, about 1840. Gordon's formula is simpler in application than those deduced by Hodgkinson, and, when properly applied, experience has shown it to be abundantly safe. The original formula is as follows for square-end columns, and should be corrected for pin or round bearings by one of the three laws above given:

$$\text{Breaking load for wrought-iron} = \frac{36{,}000 \times \text{area section}}{1 + \frac{1}{3000}\left(\frac{\text{length of column}}{\text{diameter}}\right)^2}$$

The same formula for cast-iron, using 80,000 in numerator of fraction instead of 36,000, and $\frac{1}{400}$ in denominator instead of $\frac{1}{3000}$. The constants in the numerator are intended to represent the average crushing strength of a short piece of the respective kinds of iron. Modern experiments have, however, shown that the ultimate crushing strength of American wrought-iron is much higher than that assumed in the formula—namely, 36,000 lbs. per square inch, by at least 20 or 25 per cent. In fact, so much depends upon the kind of iron, that no one constant is suitable for undeviating use. A column made from a hard iron inclined to granular, as it should be, will resist crushing better than a soft fibrous iron, or one of great tenacity, and consequently a much higher constant may be used. The following table has been computed from Gordon's formula, using 45,000 in numerator instead of 36,000, for wrought-iron—that for cast-iron remaining the same as in the original formula. It must be understood that any of the published tables for the strength of columns are purely tentative, to be modified by such light as farther experiments alone can give, and which it is hoped that the present Government Commission on the "Strength of Iron and Steel," appointed by Congress in the spring of 1875, will early institute.

TABLE SHOWING THE BREAKING STRENGTH PER SQUARE INCH OF WROUGHT AND CAST IRON COLUMNS, COMPUTED FROM GORDON'S FORMULA:

Crushing strength of wrought-iron taken at 45,000 lbs. per square inch.
" " " cast " " " 80,000 " " " "
Values given are in pounds for each square inch area.

Ratio of Length to Diameter. See page 56.	I. Breaking Load: Square Ends.		II. Breaking Load: Round Ends.		III. Breaking Load: One Round, One Square.	
	Wrought Iron.	Cast Iron.	Wrought Iron.	Cast Iron.	Wrought Iron.	Cast Iron.
10	43562	64000	14521	21333	29042	42666
15	41860	51200	13953	17067	27906	34134
20	39717	40000	13239	13333	26478	26666
25	37251	31220	12417	10407	24834	20814
30	34615	24617	11538	8206	23076	16412
35	31960	19692	10653	6564	21306	13128
40	29355	16000	9785	5334	19570	10668
45	26866	13196	8955	439	17910	8798
50	24545	11035	8182	3678	16364	7356

FIG. 16. TOP CHORD SECTIONS

FOR PIN CONNECTIONS.   RIVETED SYSTEM

The upper chord has often a similar section to that of the posts, but when not circular is usually shaped like a box, the sides of which are channel or beam irons, and the top a broad plate. The under side of such a box, when open, should be stiffened with diagonal latticing or broad batten-strips, so as to aid in the preservation of its form under the compressive strain to which it is subjected. The allowable strain per square inch on chords is governed by the same rule as that for columns; the ends being considered square, and the length of the chord the distance between two panel-points. The top chord may have simple machine-faced butt joints, or it may be made continuous in sections, the contiguous abutting surfaces being joined by fish or splice plates riveted or bolted to them. Such plates serve simply to keep the chord in position, and are not subjected to any strain whatever. Under this last arrangement, there would be attained all the advantages that can possibly be claimed for riveted work—namely, perfect continuity of material. This principle, combined with the American system, results in a structure that harmonizes theory and practice in the highest attainable degree. With some forms of compressive sections, like the Phœnix column, or the three-beam section, it is desirable, in fact necessary, that a casting be introduced to connect the several parts that cluster at the panel-points. This casting must have all its bearings machine-faced to match the faced ends of the chords and posts. In continuous box-shaped chords, the pin-holes must be reinforced with thickening plates, not only to increase pin-bearing, but also to distribute the pressure delivered to the chord at each panel-point over as much surface as possible. Further, it is advisable that the increased sectional area required at each panel-point, in approaching the centre, be placed in the *sides* of the box, as it is through the sides that the pin passes. It is not one of the least of the excellencies of the pin-connection system that the chords, posts, and tension-members may be made to unite at the centre of their several sections, and by proportioning the box chord as above this may be accomplished very fully. The advantage of a cast-iron joint box consists in the very perfect attainment of this principle, as such boxes insure an absolutely uniform distribution of pressure over the surfaces of contiguous chord sections. This principle is about as far lost sight of in riveted work as it is possible to be. In such work the chords have no stiffening along the inner edges of the vertical plates or sides to which the web system is riveted, and the increase of area is made by riveting on plates to the *upper* side of the top chord, or *lower* side of the bottom. The centre of section is not at the middle of the sides, as usually assumed, but approaches the top or bottom plates, and in large spans, where the strains are great, necessitating a large area of section (placed mostly in the above plates), the centre of section approaches the plates very rapidly. In applying the formula for posts, therefore, to such chord sections, the diameter used for determining the ratio of "length to diameter" must not be taken as equal to the side of the least circumscribing rectangle, but must be a much smaller quantity. Just what this quantity is may be ascertained by reference to special treatises on engineering, since it involves considerations too technical to introduce into a book of this character. It will be sufficient for our purpose if the reader realizes that a box or trough-shaped compression chord, having most of its metal on the upper side, is weaker than one which has the metal equally distributed among the three sides, and for the weaker chord proper allowance must be made.

In "pin-connection" chords, the pin-holes must be bored with the same care as eye-bars; the *maximum* play between pins and holes not being permitted to much exceed $\frac{1}{64}$ of an inch.

From what has been said, in describing the various systems of bridge-building in use—namely, the "riveted," the "pin," and "screw-end" connections—it will be understood how it is that the two latter can be worked very close to absolute theory, thus enabling material to be disposed in the best possible way to concentrate strains at centres of sections, and distribute them in axial lines through the various parts of the structure. Further than this, the shape of material used in designing on these systems is such that proper grades of iron are readily attainable. *The riveted system has, of necessity, so many imperfections of design, of workmanship and material, in contrast with the above, that, to obtain any thing approaching equal strength on the same specification, it should only be used with a higher factor of safety.* It is probable that this difference is not less than 20 per cent; so that when a pin bridge is called for, having a factor of *five*, a riveted bridge can not be considered as approaching the same strength unless it is proportioned with a factor of *six*. The fact that a riveted bridge is stiff, or that its deflections may be small under a test, is no evidence of strength, which last depends upon other considerations than those applying to stiffness.

The *stiffness* of a girder depends upon the average sectional area of the flanges and web, while the strength is measured solely by the *net* sectional area at any point. A girder, for example, having uniform flange areas from end to end, would be *stiffer* than one having this area only at the centre, and diminishing with the diminution of strains toward either abutment, but it would not be *stronger*. The amount of metal at the weakest point determines the strength of the girder, and since this is a matter independent of stiffness, it follows that the advocates of riveted work practice a deception on the public (perchance themselves) in pointing to the wonderful stiffness of the lattice bridge, as a triumphant refutation of the damaging criticisms made by those who have well weighed the respective merits of the various methods of bridge-building.

THE FLOORING SYSTEM.—If there is one part of a bridge more than another that can be claimed to be of supreme importance, it is the flooring system, to a careful proportioning of which more attention has been paid during the last few years than ever before. A good, stiff floor is a pretty fair criterion of the rest of the work, as well as a comfort to the travelling public. The various elements of the floor system are the cross-beams, the stringers, the connection with the trusses, the sway-bracing, and the floor-covering.

The cross-beams, often called floor or needle beams,

may be either solid rolled flange-beams, single or in pairs, or beams of lighter section deeply trussed; or, finally, riveted plate web-girders, the two last being better than the first—not that they are necessarily *stronger*, but from the great depth thereby attainable, there is less spring to

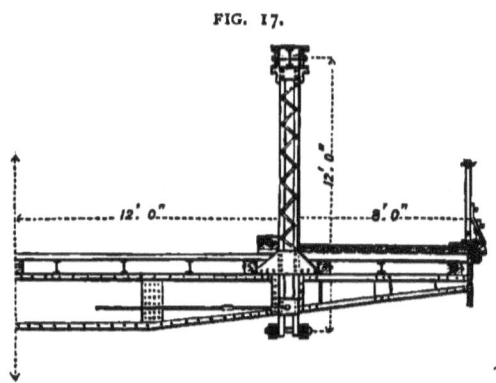

FIG. 17.

HALF SECTION PLAINFIELD BRIDGE.

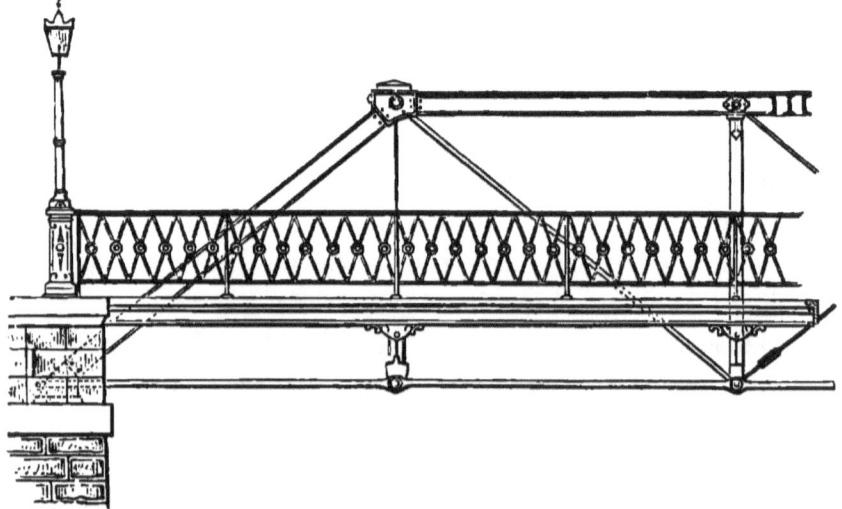

SIDE VIEW PLAINFIELD BRIDGE, 104 FEET SPAN. BY THE AUTHOR.

them under rapidly moving loads, with a proportionate gain in stiffness. In the best designs, cross-beams are located at panel-points, and they must be proportioned to carry the wheel-loads previously indicated. When sidewalks are to be carried outside of the trusses, the floor-beams of the roadway are prolonged on either side to support them, although occasionally circumstances may arise when the sidewalks must be supported by independent cantilevers bolted or riveted to the outside faces of the truss-posts. All things being considered, the compound riveted girder is probably the best form for floor-beams, because they can be made deep. A good depth for such girders in the middle is one tenth the width of roadway, but for long panels and heavy loads a still greater depth will often be found more desirable. A short distance either side of the centre, the bottom-flange may be tapered up gradually to the point of support. This form, even when not dictated by motives of economy, is very much more sightly than if the flanges are kept horizontal and parallel from end to end. The thickness of the web in such girders is usually from $\frac{1}{4}$ to $\frac{5}{16}$ of an inch, and the flanges should be so arranged as to be formed from but *two* angle-irons, the section of which must, of course, be determined by the extreme strain at centre of beam. This is a matter easily attained, since the sizes of angles vary so much that any desired area may be found in the lists of the principal manufacturers. The objections previously advanced against riveted work have least force in such girders as are above described, there being but a *single* line of rivets which unite the *solid* flanges to the web, and the number and proportion of the rivets can be computed with a fair amount of accuracy. Special attention is called to the idea of solid flanges, implied in the recommendation for using but two angle-irons, as opposed to a very common practice of using light angles, and increasing the sectional area toward the centre of the beam by riveting on plates to the angles, whereby the complexity of riveted work is introduced, which it is desirable to avoid in every instance where possible. Sufficient attention is rarely paid to the riveting, the pitch of the rivets (that is, the distance from centre to centre) being usually too great. *Thin* webs require close riveting, and the rivets should be well driven, by power if possible, since in this way alone can any reliability be placed upon the holes being well filled.

No exact rule can be given for the pitch of rivets, as it is a matter of computation in pounds, of just how much horizontal strain is delivered by the web at any given point to the flanges. As this web-strain increases toward the ends of a girder, the rivets should be placed closer as the ends are approached. The pitch will vary from 3 to 6 inches, depending upon the above considerations, and the smallest size rivet that should be used in the flanges is $\frac{3}{4}$ of an inch, which becomes $\frac{1}{16}$ greater after being driven, where the hole is properly filled. The web requires occasional stiffeners, usually two, intermediate between supports, for ordinary widths of roadway, and one at either point of support. If the web is of such thickness that the distance *between* the flange angle-irons is not greater than thirty to thirty-five times that thickness, no stiffeners will be required. Since there is no difficulty in obtaining the pieces composing a compound girder in one length between bearings, nothing has been said about joints. Should these occur, either in flange or web, pains must be taken to have splices of ample size, and a full complement of rivets, to thoroughly transmit the strength of the solid sections so united. Solid rolled beams are 10 per cent stronger than riveted beams, but are much more expensive per pound, the difference at present (1875) being 25 per cent and upward.* Such beams, in double-track roadways, from their shallowness, spring too much, throwing the trusses into an annoying vibration, to say the least, even from light passing loads, and conveying an idea of weakness, which the structures may not really possess.

The connection of the floor-beams to the trusses, for deck-bridges, is a very simple matter, as they are then directly bolted to the top chord. For through bridges, or half-deck bridges, they are either hung from the pin by means of hanger-bolts, or they are riveted or bolted to the posts. When hung from the pin, the hangers are best of the ∩ form, the legs being long enough to pass down the full depth of the floor-beam at that point, through a washer-plate (by preference of wrought-iron)

---

* Since the above was written, the price of beams has been reduced fully the amount of this difference.

on which the beam rests. The end of each leg is furnished with a nut, sometimes with a jam-nut in addition, which, when drawn up, holds the beam securely in place. Inasmuch as these hangers are short, and always feel at once the effect of the passing load, they should be of first-quality iron, and not be strained in excess of 8000 lbs. per square inch at the root of the screw-thread. They should have a flat bearing on the pin, and may be either single or in pairs. When the beams are riveted to the posts, usually between them, the connection is made by means of angle-iron brackets, one on either side of the web, and in length equal to the whole depth of the beam at the bearing, and since this attachment depends solely on the strength of the riveting, and since the riveting must be done on the ground after the work is in position, an excess of rivets should be arranged for, to compensate for the imperfections of field-riveting, which is usually more difficult to get at than in the shop, and consequently not so well done.*

The horizontal or sway bracing may consist of very light rods, if the floor is well laid, forming as it does a very effective system of bracing against lateral movement. Rods from $\frac{3}{4}$ to 1 inch round will cover all but extreme requirements, and they are attached by any convenient means to the floor-beams near their point of support. They require a screw-adjustment of some kind, turnbuckles or end-screws, in order that they may be drawn up taut. On top of the floor-beams, and lengthwise with the bridge, are laid the *stringer-beams*. These beams

* See Plainfield Bridge, page 63.

may be either of wood or iron, and are spaced from two to three and a half feet apart, depending on the character of the flooring and the loads to which the bridge is liable to be exposed. If stringers are proportioned for *wheel* loads, as has been recommended, their size is independent of their distance apart, since, however great their number, a wheel may be immediately over any one, straining it to the maximum. Where a roadway is regulated by guard-timbers, confining the wagon-tracks to a fixed position, the stringers may be made heavier immediately under the track-way, and lighter under the rest of the flooring. For wooden stringers, white or yellow pine is the best kind of timber, such varieties of timber being obtained of *straighter* grain than most any other, and consequently are peculiarly well adapted for resisting the effect of transverse strain. Stringer timbers should be inspected with greater care than is given to the floor-planks, not only on account of their position as beams, but also because floor-planks, under most circumstances, will wear out before they will rot out, while the stringers, not being exposed to the abrasion caused by horses and vehicles, become destroyed by decay, the date of such destruction being dependent on the practical knowledge of the timber inspector. Wooden stringers should be uniformly notched down on the cross-beams, which not only aid in retaining them in their position, but also insure uniformity in the level of their upper surfaces. The following table will be found convenient in determining the size of timber to be used for different panel-lengths:

PROPER SIZE STRINGERS FOR GIVEN WHEEL LOADS.

Table giving proper size of wooden stringers, for supporting different assumed wheel-loads, supposed to be concentrated in the middle of a panel, the timber being strained to 1200 lbs. per square inch.

LOADS ON ONE WHEEL.

Span or panel-length.	500 lbs.	1000 lbs.	1500 lbs.	2000 lbs.	2500 lbs.
10 feet.	2 × 8	3 × 8	3 × 9	3 × 9	3 × 10
12 "	2 × 8	3 × 8	3 × 10	3 × 12	4 × 12
14 "	2 × 9	3 × 9	3 × 11	3¼ × 12	3 × 13
18 "	2 × 10	3 × 10	3 × 12	3 × 14	4 × 14
20 "	3 × 10	3 × 12	4 × 12	4 × 14	4 × 16

Iron stringers are simply rolled I beams, of proper strength for the wheel loads, and may be had of any depth from four inches (weighing ten pounds per foot), upward. Where they rest upon the floor-girders, they should be secured by means of bolts, clips, or brackets.

* *Table giving proper size of iron stringers, for supporting different assumed wheel-loads, supposed to be concentrated in the middle of a panel, the iron being strained not over 12,000 lbs. per square inch.*

LOADS ON ONE WHEEL.

Span or panel-length.	500 lbs.	1000 lbs.	1500 lbs.	2000 lbs.	2500 lbs.
10 ft.	4" 6 lbs. p.ft.	4" 10 lbs. p.ft.	4" 10 lbs. p.ft.	5" 10 lbs. p.ft.	5" 12 lbs. p.ft.
12 "	4" 6 "	4" 10 "	4" 10 "	5" 12 "	6" 13 "
15 "	5" 10 "	5" 10 "	5" 10 "	6" 13 "	6" 13 "
18 "	6" 13 "	6" 13 "	6" 13 "	7" 18 "	7" 18 "
20 "	7" 18 "	7" 18 "	7" 18 "	7" 18 "	7" 18 "

FLOORING.—The flooring of common road-bridges usu-

* The sizes of beams recommended in the table are the nearest mercantile sizes that fulfil the requirements. This in some cases necessitates the use of beams in excess of that called for by the loads. It was thought best to use none lighter than the seven-inch beam for the twenty-feet panel-lengths, since shallower beams would be apt to spring to an undesirable extent.

ally consists of one course of plank, laid transversely to the stringers, and about three inches in thickness. Occasionally two courses are used, in which case it is a good plan to apply to the lower course some of the wood-preservative processes, and the cost of such application can be balanced by using a cheaper grade of timber than would otherwise be proper—such as spruce. If wooden stringers are used, they may also be chemically treated, when the sub-floor can be regarded as measurably permanent, the only renewals being that of the upper hardwood plank, as it becomes worn. When two courses are used, the lower one should be not less than two and a half or three inches in thickness, and the upper two inches, which last, if laid *diagonally* to the lower course, will materially stiffen the floor as a whole. The planking is spiked directly to the stringers, if of wood, with spikes having a length of about double the thickness of the plank. When two courses are used, each course should be spiked down independently. It is not necessary to spike at each stringer intersection, every other one being sufficient; but where spiked, there should be two spikes used, one at either edge of the plank. Where iron stringers are used, the simplest method of securing the floor-plank is to lay a spiking timber on either side of the roadway, and one or more between, to which the planks are fastened in the usual way. This arrangement avoids the necessity of using hook head-bolts, clinch-spikes, and other troublesome devices required if the attachment is made directly to the iron. On either side of the roadway there should be bolted to the flooring, *guard-*timbers of hard wood, with the inside edge chamfered off to make a finish. These guards should be located far enough from the trusses to prevent the wheel-hubs from striking them, and they should be raised by means of blocking, at intervals of five or six feet, about three inches, to aid the drainage, and add to their effective height. Pieces of the floor-plank, about eighteen inches long, will be found convenient for this blocking. The guard-timbers had best have lap-joints, which laps should be about twelve inches long, and secured with two bolts. Where there are sidewalks, it is desirable to have them raised above the level of the roadway, which can best be done by means of hard-wood bolster-pieces, at intervals of about four feet, laid transversely with the stringers, and of a depth equal to the desired elevation of walk. With sidewalks projecting beyond the trusses, necessitating a stiff independent railing, a rail-base should be fastened with two bolts to the ends of the bolsters, and have a projection of about three inches. This rail-base is usually from twelve to sixteen inches in width, the upper edges being neatly chamfered, and the exposed surfaces planed. On the inside edge of the bolsters, and bolted to them, next the trusses, there should be a deep guard-timber, at least twelve inches higher than the walk, and if desired, as an additional precaution, a few slats can be spiked between the sidewalk and roadway-guards, covering the otherwise open space between them, unless it happens that the roadway-plank are fitted around the posts, and carried close up to the sidewalk. With the above arrangement for supporting the sidewalk, it is necessary to lay the plank (about two inches thick) longitudinally with the bridge, spiking to the bolsters with two spikes at each intersection. It always makes the most satisfactory walk to have the planks narrow, and edged to a uniform width. They should be laid one half inch apart to form drip-spaces, and in first-class work the upper surfaces of the planks should have been planed before laying, as well as that of the rail-base and inner guard.

The planed surfaces ought to be well oiled, not alone as an inexpensive finish, but also to protect the plank in a measure from sun-cracking. The best kind of wood, beyond all question, for sidewalk plank, is yellow pine. The cornice, of $1\frac{1}{4}$-inch clear pine, is fastened to the ends of the bolster-pieces, and a bold moulding is nailed under the projection of the rail-base. A very slight expense will provide a neat scroll "drop" opposite the end of each floor-beam, which, trifling as it is, materially adds to the appearance of a bridge. The above description of the floor may be considered a standard method for the general type for road-bridges; but in important city bridges, floors should be made very much more durable than has thus far been customary in this country, except in a very few localities. It is true that durable floors, either of wood or stone paving, add vastly to the cost of a structure, increasing as it does the dead load to be carried, but in many cases it is warranted by the circumstances of heavy travel, the interruption to which through frequent repairs (as would necessarily be the case for an ordinary wooden floor) would cause great

FORMS OF WROUGHT-IRON FLOOR-PLATES.

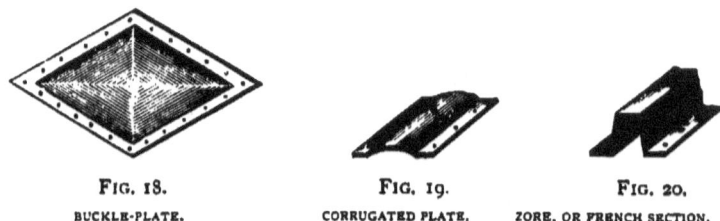

FIG. 18.
BUCKLE-PLATE.

FIG. 19.
CORRUGATED PLATE.

FIG. 20.
ZORE, OR FRENCH SECTION.

inconvenience. Any kind of paving that may be used requires an iron floor, which may be made of wrought-iron plates, $\frac{3}{16}$ to $\frac{5}{16}$ of an inch in thickness, in the form of broad corrugations laid transversely, or buckle-plates, which are rectangular plates about 3 ft. square, domed or crowned under pressure a height of three or more inches at the centre, and having flat edges on all four sides, to allow of riveting to the stringer-beams. The general appearance of these plates is that of a flattened dome. After the floor is thus formed, it must be levelled off with well-made cement concrete, to a depth of four inches and upward, to form a bottom for the paving. This concrete must be prepared with great care, as upon its excellence depends the protection of the iron plates from water, which, at the best, it is very difficult to keep from working its way through the roadway; and as floor-plates are made from comparatively thin iron, perfect immunity from rust is the price of their durability. In view of this fact, as an additional precaution, after the concrete has been levelled off, or rather crowned to the usual street regulations, and had time to harden, it is well to coat the whole surface of concrete about an inch thick with asphalt mixed with fine ashes, to add to its body, flashing it up at least six inches against all projections where it would be possible for water to trace through and get at the iron of the flooring system. On the surface thus prepared, the roadway, gutters, and sidewalks are laid as in the ordinary street, only with greater care. A proper provision for drainage must not be overlooked, and frequent spouts ought to be introduced to carry the water rapidly away, clear of the trusses. While Macadam and stone-block pavements have been used for bridge-platforms, they are enormously heavy as compared with wood, and, while more expensive in themselves than a wood-block pavement, add very largely to the general cost of all the iron-work, owing to their excessive weight. Under most any circumstances, wood blocks are the best for bridges, and if they have a good, uniform bottom to rest upon, the conditions that have caused the failure of the wood-block pavements in most of our cities are removed. Blocks four inches deep will answer all requirements for ordinary traffic, and a depth of six inches the heaviest. The blocks should not rest immediately upon the prepared floor surface, but on tarred, well-seasoned plank, one inch thick, with a thin layer of fine sand interposed between the asphalt and the plank.

BEAM-BRIDGES.—Special notice is directed to the

construction of beam-bridges, as an economical substitute for the ordinary stone arch and culverts so much in

FIG. 21. BEAM-BRIDGE—PLANK FLOOR

use throughout the country. Apart from economical considerations, they afford an increased water-way, and thus avoid the liability to disastrous overflows during sudden freshets, as is almost sure to be the case when a freshet meets with an obstruction like an arch, which, if made large enough to easily pass *extreme* floods, would become comparatively a very expensive affair. There is hardly a town or village through which a brook runs, that has not suffered more or less damage through the incapacity of arch culverts to carry off the water of an unusual freshet. Beam-bridges can very readily be carried up to spans of 25 or 30 feet, and if properly designed, and the exposed parts occasionally painted, can be regarded as durable as the old-fashioned stone arch. The flooring of such bridges may be simply plank (Fig. 21), or it may be made permanent, as before described, with iron floor-plates and paving (as in Fig. 22).

FIG. 22. BEAM-BRIDGE—NICHOLSON PAVEMENT ON BUCKLE-PLATES.

A very excellent floor is one made with brick arches turned between the beams, and laid in cement mortar, very similar to the ordinary fireproof floor (see Fig. 23).

FIG. 23. BEAM-BRIDGE—TELFORD PAVEMENT ON BRICK ARCHES.

The arches are levelled off with concrete, and the paving, or Telford, laid on the concrete surface previously coated with asphalt. For these bridges, solid rolled beams or compound plate-girders are used, spaced from 3 to 5 feet apart, with tie-rods at intervals connecting their lower flanges. The compound beams, not being restricted in depth, and costing less per pound, will usually be found the most desirable. The temptation to use thin web-plates in such girders, from motives of economy, should be avoided, as a percentage of rust must be provided for, either on account of possible neglect, or from carelessly-laid brick-work and concrete, allowing water to trace in alongside of the inaccessible plates. Before brick arches are turned, a further precaution than those named should be used, and that is to thickly coat the girders with a tar paint of some kind. Perpetuating the life of iron-work is very often simply a matter of inexpensive, preliminary precaution, which, if once realized, would be oftener put in practice than it is.

WIDTH OF ROADWAYS AND SIDEWALKS.

As to the proper width of roadways and sidewalks, where street regulations do not impose carrying the whole width of the street over the bridge, the circumstance of location is very occasional where more than two wagon-ways are necessary. Eighteen feet between the side-guard timbers are amply sufficient for all ordinary traffic, and in many cases sixteen feet will be found sufficient.

A greater width of roadway (excepting sufficient width is added for a *third* wagon-track) involves an unnecessary expenditure of money, since the bridge, being proportioned for a certain number of pounds per square foot, each unnecessary foot in width, requires just so much more material, which rapidly becomes transformed into dollars, without a particle of advantage accruing. The great difference will be found in the floor, since the cross-beams increase in weight very rapidly, as the width of the roadway increases, and the number of stringers is also increased. A rule then to determine how wide a roadway should be made is to determine the minimum width, with a margin for clearance, for one wagon-way. Then *two* or *three* times, this, according as there is a double or triple wagon-way to be accommodated, will give the distance between roadway-guards. Sidewalks, if on either side, need not be made wider than four feet in the clear; but if only one sidewalk is to be provided, having to accommodate travel in opposite directions, a width of six feet in the clear will be found sufficient.

When bridges are of such span as to necessitate a height of truss requiring overhead sway-bracing between the trusses, a clear height of from thirteen to fourteen feet above the flooring will be found to answer all but extreme requirements.

WEIGHTS OF MATERIAL, ETC.

In designing a bridge, the weight of the flooring must be first computed, and it is a fixed quantity, independent of the span, for the same width of roadways, sidewalks, and panel-lengths. It forms the principal part of the dead load in spans up to about 100 feet, and in addition to the weight of material of which it is composed, some consideration must be paid in northern climates to snow-loads, which add to the dead weight ten to fifteen pounds per square foot.

It is impossible to give a reliable rule for the dead weight of the iron and other materials entering into the construction of a bridge, depending as they do upon peculiarities of form and construction; but the following data, as far as it goes, will assist any one in determining this important preliminary in proportioning the parts of a given design. A *yard* of wrought-iron, having *one* square-inch section, weighs *ten* pounds. So that, knowing the area in inches of a given piece of iron, all that is necessary is to multiply it by ten and divide by three, to have the weight per lineal foot. A cubic inch of cast-iron will weigh ten per cent in excess of one quarter of a pound. The weight of timber varies according to its condition, whether dry or wet, a fair average being given as below:

White pine, 3 lbs. per sq. ft., B.M. 6 lbs. for 2 in. plank, 9 lbs. for 3 in.
Yellow   "   4 "   "   "   "   8 "   "   "   "   12 "   "   "
Oak   "   4½"   "   "   "   9 "   "   "   "   13½" "   "

FOR PAVEMENTS.

Wooden block, as Nicholson, 25 to 35 lbs. per square foot, according to depth of blocks.
Telford and Macadam...................130 lbs. per cubic foot.
Stone block.............................150 "   "   "   "

SUPPORTS OF PAVEMENTS.

Wrought-iron Buckle-plates, etc., depending on thickness: a square foot of one quarter inch plate-iron weighs 10 lbs.
Brick-work, when turned arches are used......120 lbs. per cubic foot.
Concrete, for levelling off ....110 to 130 "   "   "   "
Gravel......................................120 "   "   "   "
Hard asphalt................................140 "   "   "   "

THE MAINTENANCE OF IRON-WORK.

This subject has not received the degree of attention which so costly a structure as an iron bridge warrants. Too often insufficient painting is allowed to remain as the only protection for years, the fast-accumulating rust either not being noticed, or is not seen, owing to the peculiar color of the paint which may have been used. Because a bridge is an iron one, it does not imply that it requires no further care after it is once finished. When iron is *neglected*, it is only a question of time as to its final destruction. A large bar will rust out only less rapidly than a small one, or a thick plate than a thin one, and there are circumstances of location that will cause rusting to proceed with varying rapidity. It is with a view to permanence of iron structures that it is recommended in *no case* to allow of plates or parts to be used *less* than one quarter of an inch in thickness, and perhaps five sixteenths of an inch would be still more desirable as a minimum thickness. It is further advisable to have iron bridges so designed that all parts of the work should be open to inspection, and within reach of the paint-brush. When not so designed, concealed surfaces should be hermetically sealed, so that by no possibility can moisture find its way within to work a sure destruction. Town authorities should insist upon more care being exercised at the construction-works, in preparing iron for shipment, than is usually given to such matters, particularly in times of close competition, when the profit of a contractor is made up from small economies. This extra care will amply repay the very small addition to the price that it would necessitate.

At the manufactory, *each* individual piece can be examined and protected with a care impossible to exercise after the parts are all assembled in position at their final location. All new iron, as it comes from the rolling-mill, has a scale on its surface easily detached under vibration. More or less falls off while it is undergoing fabrieation into shape, but enough usually remains on, to render ineffective the paint with which it may be coated. This scale should be thoroughly removed at the shop by scraping, or with wire brushes, after which a priming coat will take hold. Some authorities recommend that before scraping off the scale, the iron should be allowed to rust slightly, as giving a better hold for the paint. In any case, the paint should be thoroughly well rubbed into the surface, and the boiled oil and turpentine with which it is mixed, and on which its value largely depends, should be of the first quality. All things considered, the mineral paints prepared from iron ores are the best priming paints, since they are inexpensive, and therefore unadulterated, which can not be said of many of the red leads (a favorite priming paint with some engineers) in the market. Before shipment, iron surfaces that have had machine-work put upon them, called bright iron, should be coated with tallow, to which a body of white-lead has been given. After a bridge has been erected, it should have at *least* two coats of tinted lead paints, care being taken that the brush reaches all the crevices about the joints. The color of the final coat or coats had better be of such a tint as will show the first indication of rust. All tints bordering on cream, buff, and different greys, answer this purpose excellently well; and as an additional advantage, these tints form a pleasing and appropriate ground for decorative effect, occasionally required for first-class city bridges. It is recommended that all iron bridges should have two additional coats of lead paint the second season after their erection, which will last several years before requiring renewal, and it would be good practice for the authorities of every county to examine their bridges systematically every spring for signs of rust, which, if discovered, should be attended to as soon as possible. In this way their bridges (if originally good ones) can be made to last forever.

THE ARCHITECTURE OF BRIDGE-BUILDING.

In the true sense of the term architecture, unadorned construction is as much a part of architecture as the more popular idea that it simply covers the art of producing pleasing effects. A man can not be a good architect before he is a good constructionist, no matter how dextrous he may be in devising graceful forms, or artistic in his selection of colors. In bridge-building, there is little room for artistic architecture, and any pleasing effect produced must grow out of consistency of design, and a thorough knowledge of the peculiarities of materials of construction and color. To an educated person, correct construction always produces a sense of satisfaction, for in it is involved the idea of proportion and appropriateness for the service to which it is put. Concealment of constructive forms, by mouldings, panels, or other devices, to suggest something else than what the construction really is, is vulgar as well as dishonest. To construct a girder bridge, and give it the *appearance* of being an arch, illustrates what is here meant by falsity in architecture,

FAIRMOUNT BRIDGE AT PHILADELPHIA, BY THE KEYSTONE BRIDGE COMPANY.

specimens of which more than one of our public parks contain. Possibly to bridges more than to any other class of public works does the Ruskinian axiom (which can not be repeated too often) apply: "Decorate the construction, but not construct decoration." Such a principle conscientiously kept in view can not but result in else than good work. Its violation results in a senseless fraud, demoralizing to the taste of the community where such violations may occur. Public works, in a certain sense, play a part in the education of a people, and their authors and builders have consequently, to that extent, a responsibility in addition to the mere utilitarian idea of endurance and safety. The ideas herein advanced are not novel ones by any means; but they can not be enforced too often, when in this boasted age of culture and civilization a community will permit the huge architectural fraud of the Fairmount Bridge over the Schuylkill at Philadelphia, and hardly yet completed. Constructively, this bridge, with its double tier of floors, spanning the Schuylkill, in a single stretch of 340 feet, is a monument to its designer and an honor to American engineering. Instead, however, of letting the enormous trusses stand in all their grandeur, depending wholly upon judicious painting and the design of the cornices and railing, etc., for their æsthetic effect, thousands of dollars have been spent in actually covering up the trusses to a great extent with sheet-iron, forming an arcade as it were of great massiveness, by arching between the posts of the trusses, the arches springing from large Roman sheet-iron capitals about *half way down* the posts! The result is that, at a little distance, the spectator beholds an arcade, without any visible means of support for a distance of 340 feet. To be thoroughly consistent, the architect (heaven save the name!) of this constructed "decoration" should have at least sanded his sheet-iron when painted, and marked out in strong lines the joints that masonry of similar forms suggests. About one mile north of this bridge, a noble structure spans the Schuylkill, the Girard Avenue Bridge, as it is called. As an *engineering* accomplishment, it stands in no comparison with the bridge at Fairmount, the spans being much smaller, and only a single roadway (of paved granite) is carried on the upper chord, it being a "deck-bridge." *Architecturally*, it is certainly one of the finest, if not the very finest, bridges in America; while in the same sense the Fairmount bridge is the worst, and probably the worst in the world. The Girard Avenue is an example of pure *decorated construction*, and the writer is aware of no place in this country where the principles for which he has been contending can be so well illustrated as in the case of these two Philadelphia bridges. A thirty-minutes' walk will carry a spectator between these two extremes of very good and very bad bridge architecture.

As before remarked, a truss-bridge presents little opportunity for architectural effect, further than what is due to correct construction, and the taste shown in the colors with which it is painted. In a through bridge,

GIRARD-AVENUE BRIDGE, PHILADELPHIA. BY PHOENIXVILLE BRIDGE WORKS.

and where the span is such as to necessitate a depth of truss requiring overhead sway-bracing, neat cornerbrackets (either of wrought or cast iron) connecting the vertical posts with the horizontal struts of the upper sway-bracing, may be appropriately introduced, since they act as knee-braces, materially stiffening the trusses against vibration. They may be made constructively useful and artistically pleasing. In those designs involving the use of cast-iron joint-boxes between the upper chord sections and posts, these boxes may be cast with neat mouldings and necks, forming capitals for the posts, in any conventional architectural forms. The effect of such caps should depend entirely on the strength of the mouldings, and not on detached leaves and pieces screwed on after the casting is finished.

When trusses terminate in vertical end-posts, there is considerable room for good effect, in making the necessary stiffening end struts or portals of such form as to embody true architectural expression. Such a design may be worked out either in cast or wrought iron with an appropriate degree of elaborateness. In doing so, however, the main lines of the portal must form an integral part of the construction, contributing to stiffness, and any appearance of brackets, arches, scroll-work, etc., *hanging* from a horizontal strut, must be avoided. The capitals of end-posts, when vertical, can be made a very prominent feature of the portal design, inasmuch as a large casting is usually required at the juncture of the end-post and top chord to accommodate the large main end-braces terminating at that point. When economy of design dictates the use of *inclined* end-posts, the portals will produce the most favorable effects, by confining the architectural effort to the expression derived from the simple bracing-bars. An arch-portal of angle or T iron, with the spandrils filled in with lattice-work, or broken up into triangles with bracing-bars, is simple and expressive, and exceedingly appropriate. The lattice intersections can be ornamented by small rosettes or bosses, and the two halves of the portal-arches can be united properly with a half-circle or other form of centre-piece, while at the springing, where they are bolted to the sides of the end-posts, a neatly-designed bracket or shoe will not be out of character. It is exceedingly difficult to design the portals for inclined end-posts so as to look well, since they are viewed obliquely, and it will be found in such cases that simplicity of design growing out of an agreeable arrangement of constructive necessities will always give the best results. The appearance of a roadway-bridge having sidewalks is very much enhanced, and at a very small cost, by neatly-designed railings, with a deeply-moulded fascia-board, to which may be added scroll-drops opposite the ends of the floor-beams.

It is not necessary that such railings should be expensive, a light lattice railing of wrought-iron, with one or more intersections, with or without rosettes, always looking well and harmonizing with the constructive character of a truss. The cheap gas-pipe railing is so positively ugly that its use ought to be banished to those country districts where it is rarely seen. Well-designed newel-posts, lamp-posts, and brackets are features of a design where a cultivated taste may be exercised, and form no small part of the prominent accessories of public works of this character. This matter of treating bridge constructions as architectural works, in the true sense of that term, deserves the most thoughtful consideration of engineers and committees, as bridges nearly always form prominent objects of observation in cities and towns, particularly when across large watercourses. They are seen by every one, and therefore in those portions and surroundings capable of æsthetic treatment, some regard should be paid to appearances. A plain four-walled building—as a court-house for example—might answer every requirement for public purposes, but the demands of modern civilization require that a large expenditure must be made for what is called "architectural effect," in order that a certain gratification may be derived by the community where it occurs, springing from the contemplation of pleasing forms. Nothing has been said about masonry design, as in these pages we are simply dealing with the superstructure, but as the masonry forms part of a bridge design when taken as a whole, the form of piers, abutments, character of masonry, coping, etc., it must not be forgotten, leave abundant room in many cases for the exercise of correct æsthetic treatment. There are very few who can not appreciate a well-proportioned pier, with its ice-breaker, heavy coping and belting courses, well-laid, rock-faced work, and chisel-drafted corners.

TESTING.

As to the utility of testing individual pieces of work during manufacture, opinions differ, but it is unquestionably a wise procedure, in the case of welds in main tension bars, as imperfections of workmanship and material (if any exist), undiscoverable by the eye, will be very apt to be developed under strain. To avoid injury, it is advisable that this proving should not be carried beyond say nine tenths of the elastic limit; thus a bar with an elastic limit of 20,000 lbs. per square inch should not be tested much beyond 18,000 lbs. After erection, all bridges should be tested with loads approaching as near as possible the maximum loads for which they were designed. Railroad-bridges are very readily tested, but highway-bridges can only be tested at considerable expense. Pig-iron, or paving-blocks when convenient, are probably the best artificial loads that can be used, as they are readily handled and distributed. An excellent, though expensive, method of testing, and one of universal application, is to distribute gravel in a uniform layer over the whole area of roadway, and of such thickness as to equal the load which the bridge was designed to carry. Inasmuch as the weight of gravel and earth varies according to locality and degree of moisture when excavated, before a proposed test is made, a cubic foot of the testing material must be weighed to determine the proper thickness to be put on the bridge. In order to judge the result properly, means must be used to measure the deflection of the structure undergoing the test, which may be done by observations with a levelling instrument, or when convenient (as in most cases) by planting a pole or measuring-rod alongside of the span at the centre, as follows: After the bridge is completely finished, and come to a natural bearing under its own dead load, observe the position of any part, say the lower chord at centre, with reference to the position of the instrument or measuring-pole. Then apply the test load, and measure the amount of deflection caused thereby. Remove the test, and observe again how near the bridge returns to its first position. This it will do if the bridge is well built, *less* a small fraction due to that peculiar quality of wrought-iron which is called "permanent set," which takes place under comparatively very small strains. The set here spoken of must not be confounded with that taking place after the elastic limit is reached, but simply means that the various parts of the bridge have come to a working bearing. If the test load is now applied for the *second* time, as it always should be, it would be found that the deflection would be precisely the same as it was before, under the first test, and so also the amount of recovery after the load was removed. To make a real test, this second application of the load, with accompanying observations, should not be omitted. To illustrate: suppose, at the first loading, the deflection was two inches, on its removal the span recovered itself within one eighth of an inch. This proportion of the deflection is permanent, due to the span coming to its bearing, and will forever exist. The *second* loading would now produce a deflection of $1\frac{7}{8}$ inches instead of 2 inches, as at first, the total $1\frac{7}{8}$ plus $\frac{1}{8}$ of an inch being precisely the amount of the original deflection. Upon removal of the load, the recovery would be $1\frac{7}{8}$ inches, the same as before. Bridges ought always to be built with a camber or upward curvature, which camber at a minimum should not be less than the deflection caused by a maximum loading. Beyond this the amount is purely a matter of taste with the designer, it having nothing whatever to do with the strength of the work.*

BRIDGE-LETTINGS.

In the matter of "Lettings," it frequently happens, that parties with the best intentions make mistakes against themselves in their award, simply from ignorance of what they really do want, and by so doing are apt to work an injustice toward competing parties, that is provocative of suspicion and ill-feeling all around. With a view to aid in a clear understanding of how bridge-lettings should be conducted, in order to secure the best results at the least cost, the following forms of invitation and specification have been prepared, in the hope that they will save well-meaning committeemen much perplexity.

While the forms recommended are brief in expression, they cover all the salient points necessary for a fair competition. The specifications are general, and should

---

* All bridges, besides being tested for deflection, under a dead load, should be tested to see that they are measurably free from vertical and lateral vibration, owing to lack of counter and horizontal bracing. The best test for this purpose, is to have a couple of heavily loaded carts driven rapidly back and forth over the roadway.

CHESTER PARK BRIDGE, HAMILTON COUNTY, OHIO. By *Cincinnati Bridge Company.*
134 feet Span ; 18 feet Roadway ; 5 feet Sidewalk ; 100 pounds per square foot ; Factor of 4 for safety.

be made so, as the best work is obtained by permitting bridge-builders to have full latitude of design, under no other restriction than that of requirements and material. These should be made so clear that no refuge for evasion may be found under technicalities. To make a just comparison of prices, competing parties must estimate upon precisely the same basis, or endless confusion will result in any effort to make a fair canvass of tenders. It is recommended, in all cases of a bridge-letting, to call in the services of an expert—not simply a general engineer, but one *familiar* with the science and practice of bridge-building, for the purpose of examining the strain-sheets submitted with the tenders, and comparing them with the specifications on which bids were taken. His services should be continued throughout the building of the bridge, the work on which, however, should not be commenced before all *detail* drawings have been made by the contractor, and submitted to the expert for criticism and approval. If it is inconvenient to employ such an inspector through the continuance of the work, he should be called in at its completion, to make a thorough examination as to the material and execution, in accordance with the contract and specification. A suggestion was made in a report[*] to the American Society of Civil Engineers

---

[*] Report on the "Means of Averting Bridge Accidents," by James B. Eads; C. Shaler Smith, of St. Louis; I. M. St. John, of Louisville; Thomas C. Clarke, of Philadelphia; James Owen, Newark, N. J.; Alf. P. Boller, Octave Chanute, and Charles Macdonald, New-York; Julius W. Adams, of Brooklyn, and Theodore G. Ellis, of Hartford, Ct.—*Transactions American Society Civil Engineers*, 1875.

on the subject of " Bridge Accidents," which deserves the very serious consideration of town authorities. It was to the effect that every bridge built should have a tablet fixed upon it in a conspicuous place, on which should be inscribed the name of the builder, the expert inspector, the names of the committee or corporation officers under whom built, the load for which it was proportioned to carry, with factor of safety and date of erection. Such a method of procedure tends to fasten responsibility, which is a powerful incentive to honest, conscientious work, and if every State passed a law covering the above suggestion, there would in a short time be a surprising improvement in the design and construction of highway bridges, although that improvement would be accompanied with an increased cost.

It will be noticed, in the last clause of the form for "Invitation," bidders are requested to be present at the opening of the bids, and hearing them read. This is simple justice; and when one considers the time required to make plans and estimates, even for a small piece of work, to say nothing of the expenditure of money incident thereto, with probable travelling expenses in addition, no fair-minded man can object to rendering at least what satisfaction may be derived from the public opening of tenders. Bids secretly opened always lead, whether justly or unjustly, to the suspicion of unfair practices, an imputation that can be readily removed by the method of publicity suggested, a method which can be objected to by no one, unless those whose mode of doing business seeks darkness rather than light.

PROPOSED FORM OF INVITATION TO " BRIDGE-BUILDERS.

.............., 187

The undersigned committee of .... will meet at .... at .... o'clock, on the .... day of ......, for the purpose of receiving plans and proposals for the furnishing of all material, the construction and erection of a wrought-iron bridge over ...., agreeably to the specifictions hereto annexed. Parties tendering must furnish a clearly made-out strain-sheet of their design, with the data on which it was computed, and showing also the areas of material proposed to be given to each part. Bidders are requested to be present on the above occasion, when all the proposals will be opened and read in their presence. The right to reject any or all bids is reserved.

Signed by the Committee,

....

PROPOSED FORM OF STANDARD SPECIFICATION FOR ORDINARY HIGHWAY BRIDGES, WHEN INVITING TENDERS.

GENERAL DESCRIPTION.—The bridge will be a (through or deck) bridge, consisting of .... spans, and will have a roadway of .... feet between guards, with .... sidewalks of .... feet in the clear each. Sidewalks to be raised .... inches above level of roadway.

The distance from grade to bed of stream (or from

grade to grade of roads crossing each other) is .... feet. From grade to highest water is .... feet, and the centre line of the bridges makes an angle of .... degrees (to the right or left), with the face of abutment or piers.

LOADS TO BE CARRIED.—The bridge must be proportioned to carry, in addition to its own weight, ... lbs. per square foot (see table, page 16) of moving load, starting at one end, and moving over until the whole span is covered, in addition to which the flooring system must be proportioned to carry .... tons (of 2000 pounds) on each pair of wheels for each wagon-way, and due consideration must be given to the effect of this concentrated loading upon the posts and tension-braces of the trusses. The stringer-beams and floor-beams (to be wood or iron, as desired).

FACTOR OF SAFETY.—Under the above loading, the factor of safety referred to *ultimate* strength, shall be for the chords (4 or 5), for the web-system 5, and for all parts of the floor system (5 or 6).

MATERIAL.—The *wrought-iron* used shall be of that quality best suited to the purpose, the test for bars being that pieces cut therefrom shall be capable of being bent cold without fracture, until the two sides of the bend shall approach each each other within the thickness of the bar. No iron in small bars to be used with an ultimate strength of *less* than 55,000 pounds per square inch, or an elastic limit of less than 24,000 pounds per square inch. *Castings* must have a clean skin, free from holes or cinder and expose when broken a fine-grained grey fracture. *Lumber* must be of a good, merchantable quality, sound and free from black or loose knots and wind-shakes, and not have sap on more than three corners for planks, or on two for stringer-timbers, or wany edges on more than one corner. For roadway plank the lumber will be of ...... three inches thick, for sidewalk plank, two inches thick of ......, and for stringers ...... pine will be required.

CONSTRUCTION.—In pin-connection designs, the *pins* must be carefully turned to match the holes of the several parts of the trusses through which they pass, with a minimum play of a scant $\frac{1}{32}$ of an inch, and in diameter must not be less than $\frac{8}{10}$ the width of the largest bars they connect, if of flat iron, or if the bars are of square iron the diameter must not be less than $1\frac{3}{4}$ times the side of the largest square. The heads of eye-bars must have at least 50 per cent of effective section more than in the body of the bar. The bearing surfaces of the compression members on the pins must be effectively reinforced, so that the minimum thickness in inches of such surface will not be less than the result derived by dividing the maximum strain as shown on the strain-sheet in pounds, by 12,000 times the diameter of the pin. All bearing surfaces must be machine-faced, and any discrepancy in length between all parts in the same panel must not exceed $\frac{1}{64}$ of an inch. Where *rivets* are used, serving to *transmit* strain, and not simply for the purpose of securing parts in *position*, they should be proportioned as to number and size by considering the *work-*

ing value of each rivet to be equal to its diameter, multiplied by 12,000 pounds, multiplied by the thickness of the thinnest plate. The plates and angle-bars subject to *tension*, under such riveted construction, must have an allowance made up for the area cut out by the rivet-holes.

Pin-connection work and solid section iron will be considered to have an advantage of 10 to 20 per cent over and above riveted or compound work. The spacing of the rivets must not exceed five inches between centres, and in the flanges of plate-girders this pitch must be reduced as the ends are approached, according to the value obtained by the above rule for the proportioning of rivets. Before shipment, all iron must have a thorough coating of mineral paint, well rubbed in, and all bright work must be protected with white-lead and tallow.

To the above specifications must be added the degree of finish required, such as the painting after erection, the manner in which sidewalk is to be laid, whether the plank is to be planed, etc.; also the kind of railing desired, whether plain or ornamental, and proposed arrangements for lighting.

HIGHWAY AND RAILROAD BRIDGE, 250 FEET SPANS, ST. JOSEPH, MO. BY THE DETROIT BRIDGE AND IRON COMPANY.

PART II.

THE STRAINS IN GIRDERS AND SIMPLE TRUSSES.

ALL questions involved in the consideration of this subject resolve themselves into mere questions of *leverage*, of greater or less complexity. It is by means of the law of the lever that we are enabled to determine precisely what portion of a given weight resting on a beam is sustained by either point of support, which is the first thing to be determined before the strains can be computed. The law is simply this: "The weights balancing each other at either end of a beam or lever over any point, are to each other inversely as their distances (called lever-arms) from the point or fulcrum." For ex-

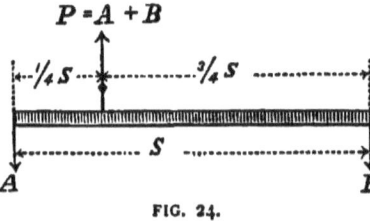

FIG. 24.

ample, supposing we have a beam held up as in Fig. 24, with a weight at either end, the point of support being to one side of the centre, say at ¼ the length of the lever from one end. Then, in order that the lever be balanced, the weight at B must be ¼ the sum of A and B, and that at A, ¾ that sum; for always B multiplied by ¾ S must equal A multiplied by ¼ S, and the sustaining force P must, of course, equal the sum of A and B. For example, suppose P, or A plus B, is 12 tons, and the span S is 20 ft. For equilibrium, the proportion of the 12 tons at A is in excess of that at B, precisely in the proportion that the lever of B exceeds that of A— in this case, 3 times. A, then, must be 9 tons, and B 3 tons, and ¼ S multiplied by 9 equals 45, being the same as ¾ S multiplied by 3. Again, supposing that there is but *one* weight, and *two* points of support, as in the

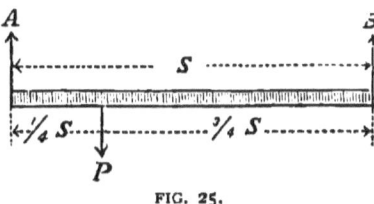

FIG. 25.

figure, the condition is precisely the same as before, only reversed, and, according to the law of the lever, we find that for equilibrium a force must be applied to A equal to ¾ of P, and at B equal to ¼ P. This last example is precisely the same case as that of a beam or truss of any kind, only A and B are now called the *reactions* of the abutments, the sum of which must always be equal to the weight or weights causing them. In order then to know just how much of the weight at any point of a beam is supported by either abutment, all that is necessary to be done is to multiply the shorter or longer segment into which its centre of gravity divides the beam (according to the above law) by the weight, and then divide by the product by the sum of the segments, which is, of course, the same as the span. For example, sup-

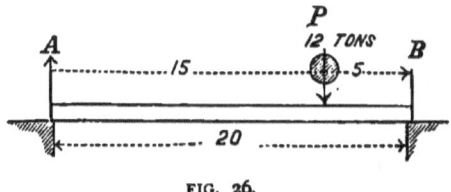

FIG. 26.

pose we have a beam A B (Fig. 26) 20 ft. long, and there is a weight of 12 tons, ¼ the distance from B, or 5 ft. Then each abutment supports or "reacts" a certain amount of this weight proportional to its distance from either end, the sum of these reactions being equal to the weight. A supports or reacts according to the rule: $\frac{12 \text{ tons} \times 5 \text{ ft.}}{15 \text{ ft.} + 5 \text{ ft.}} = 3$ tons; and B supports $\frac{12 \text{ tons} \times 15 \text{ ft.}}{15 \text{ ft.} + 5 \text{ ft.}} = 9$ tons. Adding these two upward reactions, there results a total of 12 tons, the same as the whole load at P acting downward. Any number of weights are to be treated in the same way, the sums of their separate reactions being the total reactions or weight supported at each abutment Any weight or force multiplied by the leverage at which it acts is called the *moment* of that weight or force. The leverage or lever-arm of any force is the perpendicular distance let fall from the point around which its moment is taken (or the "fulcrum") upon the direction of the force. Thus if we have a force P (Fig. 27), and the ful-

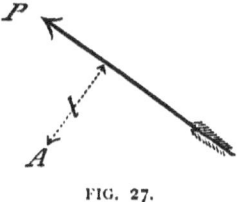

FIG. 27.

crum about which it acts is A, then l is the lever-arm of that force, and P multiplied by l the moment. Since the tendency of a force acting with a lever is to produce motion, and it being evident that all the forces acting at any given point of a beam or truss can not act in the same direction, it follows, if equilibrium is to be maintained, the sum of all tendencies to move in one direction must equal those in the opposite direction, or their algebraic sum be zero.

The ordinary crowbar (Fig. 28) is a familiar every-

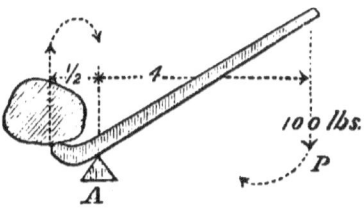

FIG. 28.

day example of the "principle of moments" above explained. Suppose a man presses down with a force of 100 lbs., distant 4 ft. from the fulcrum A. The *moment* of this pressure is 100 lbs. multiplied by 4 ft., or 400 feet-pounds, as it is called, and it acts, with reference to the fulcrum, toward the *left*. The weight that will just balance this moment acts toward the *right*, with a lever of 6 inches, or one half a foot; and since the moment of this weight must equal the moment of the pressure, the weight itself must be 800 lbs. For 800 lbs. multiplied by one half of a foot equals 100 lbs. multiplied by 4 ft. To absolutely move or destroy the equilibrium of a weight of 800 lbs. circumstanced as above would require the man to just exceed a pressure of 800 lbs., barring the resistance due to friction.

In any beam or truss, there are two sets, as it were, of forces in action, called exterior and interior forces; one tending to break the beam through bending, and the other tending to resist breakage. The former are derived from the weight of the beam and the loads placed upon it, and the other from the resistance of the material, in which is involved the form of cross-section. When a beam is bent by the imposition of a load, it is accompanied with a pulling apart of the fibres on the convex side, and a crowding together of those on the concave side. The one signifies *tension*, and the other *compression*, and in passing from one extreme to the other, there must necessarily be a set of fibres without strain. Where these unstrained fibres occur is called the *neutral axis* of the beam, and its position is, in all cases when the load is vertical, in the centre of gravity of the beam-section. The annexed illustrations (Figs. 29 and 30) show in an exaggerated way this extension

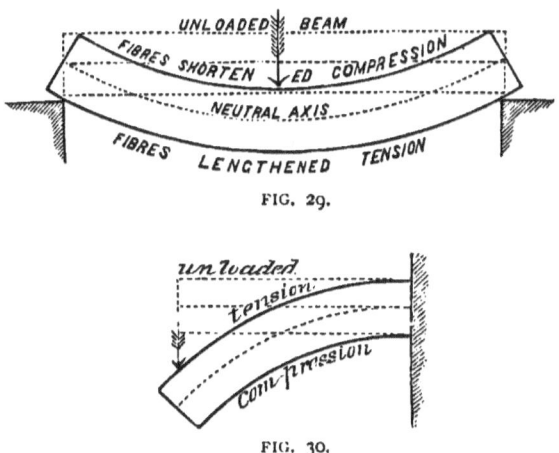

FIG. 29.

FIG. 30.

and shortening of the fibres, and it will be noticed that the fibres lengthen or shorten proportionately to their distance from the neutral axis. The relative intensity of the strain is also measured by the relative changes in length of the fibres. At the neutral axis, the fibres being unchanged in length, there is no strain; but on the exterior surfaces, the top and bottom of the beam, the fibres are lengthened or shortened a maximum amount, and the strain is there a maximum. In further illustration of this principle, suppose there is a rectangular beam (Fig. 31) of which A B C D represents a side view, with the neutral axis M N passing through the centre of gravity. When the beam is loaded with a weight W, it will deflect, due to the shortening or compressing of the fibres on the upper surface, and lengthening those on the lower, as before explained. Let ab and $a'b'$ represent the extreme changes in length of the fibres on the outer surfaces. Then, since the strain at centre is nothing, if

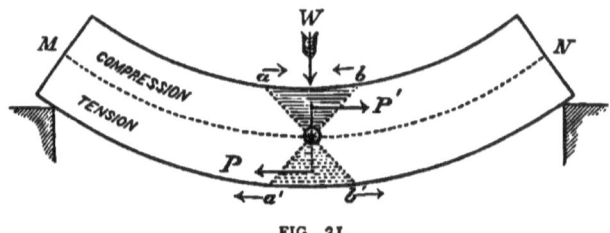

FIG. 31.

we draw two triangles either way from the centre to the points of extreme strain, the strain on any fibre will be represented by the length intercepted by the sides of the triangles aO and bO and a'O and b'O.

Summing the changes of length of all the fibres in either triangle, there results the representation of the total amount of the tensile and compressive strains, or, what is the same thing, the sum of the strains may be represented by the *areas* of the triangles, their *mean effect* taking place at the fibres of mean length, or the centres of gravity of the triangles, which is one third their height from their bases, or two thirds the distance above or below the neutral axis. This mean effect is represented in the figure by the forces P and P'. These two forces, acting in opposite directions, and parallel to each other, constitute what is called a *couple*, their leverage of action being their distance apart, which lever is also the *effective depth of the beam*. To determine the resistance of a rectangular cross-section, let C equal stress on outside fibre represented by ab or $a'b'$, $d =$ depth of beam, and let the width of beam be taken as unity. Then from what has preceded we have the average force P or P' (equal to the areas of either triangle or $\frac{1}{2} C \times \frac{1}{2} d = P$) multiplied by the leverage of action, or the distances apart of the centres of gravity of the triangles. That is to say, $P = \frac{1}{2} C \times \frac{1}{2} d \times \frac{2}{3} d = \frac{1}{6} C d^2$. For any breadth b other than unity, this expression becomes

$$P = \frac{b d^2}{6} C = \frac{\text{area of cross-section}}{6} \times d\, C \dots \dots \dots (1)$$

This is a general expression for the resistance of any beam having a rectangular cross-section, and is called the *moment of resistance* of the cross-section (usually designated by the letter R). When this value equals that due to the weight multiplied by its leverage of action, called the *moment of rupture*, or M, there is perfect equilibrium between the rupturing and resisting forces, or, in algebraic expression, "M" $=$ "R."

The constant C is called the *modulus of rupture*, and were it not for certain discrepancies that occur in the resistance of material when subjected to *direct* compression or tension, and to *cross breaking*, its value would be given experimentally by the force necessary to tear apart or compress a bar of a given material. It is unnecessary in this place to point out what these discrepancies are, but simply 'the fact. Professor Rankine recommends that the value of C for any kind of material be determined by taking eighteen times the force necessary to break with a centre load a bar one inch square, placed on supports one foot apart.* Bearing in mind the principle of the equality of moments of rupture and resistance necessary for perfect equilibrium, as previously explained, the following application to beams differently circumstanced will cover the requirements of ordinary practice.

Beam loaded at one end, fastened at the other. Maximum moment of rupture occurs at point of support. The lever which produces this is the length of the beam, or *l*.

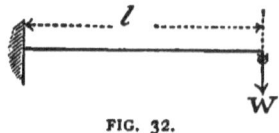

FIG. 32.

$$M^{max} = W\,l = R \text{ and } W = \frac{R}{l} \quad\quad\quad\quad\quad\quad\quad\quad (2)$$

Beam supported at one end, and uniformly loaded with *w* units per foot; *wl* will be therefore the total load, the centre of gravity of which is in the middle of beam, or leverage of action to produce mean moment of rupture is ½ *l*.

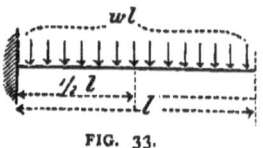

FIG. 33.

$$M^{max} = wl \times \frac{l}{2} = \frac{wl^2}{2} = R \text{ and } w = \frac{2R}{l^2} \quad\quad\quad (3)$$

Beam loaded at centre with W, and supported at both ends, length *l*. Leverage of action ¼ *l* for the reaction of either abutment, the fulcrum being immediately under the weight.

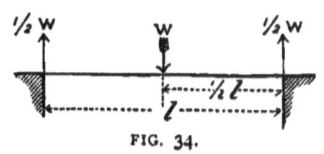

FIG. 34.

$$M^{max} = \tfrac{1}{2} W \times \tfrac{1}{2} l = \frac{Wl}{4} = R \text{ and } W = \frac{4R}{l} \quad\quad\quad\quad (4)$$

Beam uniformly loaded with *w* per unit of length, giving *wl* for total load—supported at both ends. Maximum moment under centre gravity of load, lever ¼ *l*. Reaction of either abutment ½ the whole load.

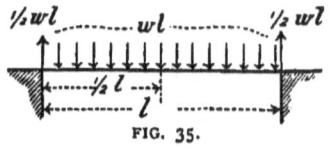

FIG. 35.

$$M^{max} = \tfrac{1}{2} wl \times \tfrac{1}{2} l \text{ less } \tfrac{1}{2} wl \times \tfrac{1}{4} l = \frac{wl^2}{4} - \frac{wl^2}{8} = \frac{wl^2}{8} = R \text{ and}$$

$$w = \frac{8R}{l^2} \quad\quad\quad\quad\quad\quad\quad\quad\quad\quad\quad\quad\quad\quad\quad\quad\quad (5)$$

* The value eighteen times the breaking force used in determining constant C is derivable as follows—see Fig. 34:

Let W = breaking weight at centre of bar.
C = required constant or modulus of rupture. (*Continued on page 106.*)

It will be noticed this last expression is obtained by subtracting that portion of the load between the abutment and centre that acts in a *contrariwise* direction to the reaction of the abutment.

If the loads are placed in any other position, or are only partial, M can always be found by first finding the reaction of either abutment (page 101), and multiplying that reaction by the distance from the abutment to the point where M is wanted. The reaction being *upward*, if there are any weights (which act *downward*) between the abutment and point of desired M, they must be multiplied into the leverages with which they act around that point, and their *sum deducted* from the product of the reaction and its leverage before found. This is the principle that had to be applied to the circumstance of loading shown in Fig. 35. As an example of the application of these formulæ: suppose in all cases the material is a pine stick $10'' \times 10'' \times 10$ feet or 120 inches long. We require to know the *breaking load* under each condition of loading, C being 7000 lbs. See formula (1):

No. 2. $W = \dfrac{R}{l} = \dfrac{7000 \times 100 \times 10}{6 \times 120} = 9722$ lbs.—hung at one end.

No. 3. $w = \dfrac{2R}{l^2} = \dfrac{2 \times 7000 \times 100 \times 10}{6 \times 14,400} = 162$ lbs. per lineal inch $= 19,440$ lbs. uniformly distributed.

No. 4. $W = \dfrac{4R}{l} = \dfrac{4 \times 7000 \times 100 \times 10}{6 \times 120} = 38,888$ lbs.—supported in middle.

No. 5. $w = \dfrac{8R}{l^2} = \dfrac{8 \times 7000 \times 100 \times 10}{6 \times 14,400} = 648$ lbs. per lineal inch $= 77,760$ lbs. uniformly distributed.

The bar being one foot long between bearings, and *one inch* square, we have the moment due to external forces $\frac{1}{2} W \times \frac{1}{2}$ span $= 3 W$.

And the moment due to internal forces $R = \dfrac{b d^2}{6} C = \frac{1}{6} C$.

Since M must equal R, we have

$\frac{1}{6} C = 3 W$; or $C = 18 W$.

Now, assuming a safety factor of five, the *safe load* to which the above stick should be subjected would be:

One end fixed, the other free; weight at free end. .... 1944 lbs.
One end fixed, the other free; weight distributed..... 3888 lbs.
Both ends supported; weight concentrated at middle. 7777 lbs.
Both ends supported; weight uniformly distributed...15,555 lbs.

It will be noticed from this example that, taking the first case as having a strength of *one*, with the second condition of loading and support, the stick will sustain *twice* as much, with the third *four* times as much, and with the fourth condition *eight* times as much. The third and fourth conditions are those that apply to the longitudinal stringer-beams of a bridge, and from formulas 4 and 5 has been computed the following table for different spans or panel-lengths and depths of stringers, the thickness being for a unit of *one inch*. The modulus of rupture C for pine has been taken at 8000 lbs. with a factor of safety of *six*.

TABLE GIVING A SAFE CENTRE WORKING LOAD IN POUNDS FOR ANY DEPTH OF PINE STRINGER AND A UNIFORM WIDTH OF ONE INCH.

Depth in Inches.	Clear Span in Feet.							
	6 feet.	8 feet.	10 feet.	12 feet.	14 feet.	16 feet.	18 feet.	20 feet.
6	443	333	266	...	...	...	...	...
7	602	454	363	302	...	...	...	...
8	787	593	474	395	338	...	...	...
9	996	750	600	500	428	375	...	...
10	...	927	741	617	529	463	411	...
12	...	...	1067	888	761	666	591	527
14	...	...	...	1209	1036	907	805	717
16	...	...	...	...	1354	1185	1052	938

For *safe, uniformly distributed* loads, double the loads given in the table.

To use the above table, the weight to be carried in the centre of a given span is first determined, and then select any depth for the beam, and follow along the horizontal line until below the span at top of column. The number there found will be the safe load in pounds for a beam of the given depth and one inch thick. Divide the weight to be carried by the number of pounds found from the table, as above, and the result will be the width in inches required for the beam. Thus, for example, it is required to know how thick a piece of timber should be that is 10 inches deep, spanning 12 feet to carry 3000 lbs. hung in the middle, or, what is the same thing, 6000 lbs. uniformly distributed. Opposite 10 in the first column and below 12 in the fifth column, we find 617 lbs., the safe load for one inch thick. Dividing 3000 by 617, we find the timber should be a shade less than 5 inches thick. The following table is given as showing judicious sizes for the wooden stringer-beams for the various classes of bridges, and for varying panel-lengths. In judging this table, it is to be considered that the standard wheel loads recommended in Part I. are extreme, and therefore very occasional, so that a much lower factor can safely be used for such loads. Under these circumstances, if the stringers are of good timber, they can safely be proportioned for a working stress of 1500 lbs. per square inch.

Span or panel-length.	Size Stringers for City Bridges.	Size Stringers for Town Bridges.	Size Stringers for County Bridges.
8 feet..................	3 × 10	3 × 10	3 × 9
10 "	4 × 10	4 × 10	3 × 10
12 "	4 × 12	3½ × 12	3 × 11
14 "	4 × 13	4 × 12	3 × 12
16 "	4 × 14	4 × 13	4 × 12
18 "	4 × 15	4 × 14	4 × 13
20 "	4 × 16	4 × 15	4 × 14

Thus far we have been dealing with rectangular cross-sections; but bearing in mind the explanation made as to the stresses on the different fibres of a beam with reference to the neutral axis, it will be at once seen how wasteful it is to have so much material near the neutral axis, where it is of so little service. If the material were so disposed as to be principally in the upper and lower portions of the beam, the strength of the beam would be largely added to. With wood, other than a rectangular section is evidently out of the question; but in iron, the true form for the most economical distribution of material is a necessity in practical construction, and is readily attained by concentrating most of the metal in the upper and lower portions or the " flanges," the stem or web being just stout enough to properly unite them, and to resist the tendency of one part of the beam to slide *vertically* or *horizontally* past the other under the direct action of the load, called the *shearing* tendency. For example, the accompanying cut, Fig. 36, represents the vertical shearing tendency of a load, which is least at the centre and greatest at the abutments, as each section either side of centre must take up the shear of each preceding one. Solid rolled beams are manufactured in this country from 4 inches deep, with 3-inch section, to

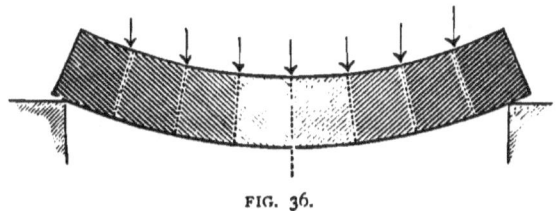

FIG. 36.

15 inches deep, with 20-inch section. Their ordinary length up to and including the 9-inch beam is, 30 feet. The beams exceeding 9 inches have an ordinary length of from 20 to 25 feet, according to weight. Beams are often rolled beyond the commercial ordinary length; but the cost of extra lengths increases very rapidly with such excess.

To determine the "Moment of Resistance" of flanged beam sections, we must consider first the resistance due to the rectangular web, and, secondly, that due to the flanges. The resistance due to the web portion has already been shown to be equal to one sixth of its area multiplied by its height, being the same as a rectangular section. That of the flanges is the area of either one multiplied by the distance apart of their centres of gravity, which, when added to the resistance of the web, gives the total resistance of the section. The web should not be taken the whole depth of the beam, but only from flange to flange. Thus, suppose we want to know "R" for the beam proportioned as in Fig. 37:

MOMENT OF RESISTANCE—FLANGE-SECTIONS. 111

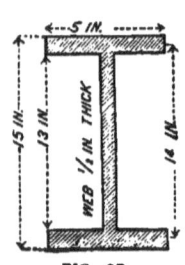

FIG. 37.

$\frac{1}{6}$ area web $= \frac{13" \times \frac{1}{2}"}{6} = \frac{6\frac{1}{2}"}{6}$ multiplied by 13" equal 14.083

Area flange = 5" x 1" = 5" multiplied by 14" equal 70.000

Resistance of section "R".................. 84.083

The quantity thus obtained has only to do with the *shape* of section; the efficiency to do work being dependent on the quality of the material. R must therefore be multiplied by a coefficient expressing this quality before the strength of the beam becomes known. For wrought-iron, this coefficient, within safe limits, varies from 10,000 to 15,000 lbs. per square inch, depending upon the requirements of any given specification. The above process for obtaining the value of R varies so fractionally from absolute truth that the refinement of calculation to obtain mathematical exactness is entirely unnecessary, while the ease of its application is so great that but a few moments of the simplest arithmetical processes are all that is required to compute the resisting value of any beam of the usual patterns.

The formulæ already given for different circumstances of loading, page 105, may be divided by the assumed maximum strain per square inch allowed on the iron, which amounts to the same thing as multiplying R by the same quantity, and is the most convenient way of introducing the above coefficient. As an example in applying the above principles for determining the proper size of beam for any given load, let us take the condition of loading given by equation 5, page 105. Let the load to be carried be 40,000 pounds, uniformly distributed, and the maximum allowable strain be 10,000 lbs. per square inch; span, 15 feet, or 180 inches. Then formula 5 would read:

$$M^{max} = \frac{40{,}000 \text{ lbs. multiplied by } 180 \text{ inches}}{8 \text{ multiplied by } 10{,}000 \text{ lbs. per sq. in.}} = R = 90$$

A beam must therefore be designed having this value of R, precisely as described on page 111. It will be noticed that the section there computed falls a little short of a moment of 90, which would be attained by increasing the flange areas ten per cent. Since each beam-section has its own value of R, the following table gives this value for all shapes of "Phœnix" beams, and is about the same for the same sizes of other makers:

TABLE GIVING THE VALUE OF R FOR ALL SECTIONS OF AMERICAN BEAMS.

Total depth in inches.	Weight per foot.	Area of one flange.	Distance between centres of flanges.	Area of stem.	Depth of stem.	Moment of resistance, R.
15	66¾	6.10	13.80	7.80	11.875	102.20
15	50	4.312	14.04	6.375	12.750	75.44
12	56¾	5.755	10.92	5.49	9.250	72.85
12	41⅞	3.790	11.16	4.92	10.000	51.48
10½	35	3.380	9.74	3.74	8.625	38.96
9	50	5.50	7.90	4.00	6.375	48.70
9	28	2.78	8.30	2.84	7.000	27.00
9	23½	2.37	8.38	2.26	7.250	22.88
8	21¾	2.035	7.42	2.43	6.500	18.11
7	18¼	1.80	6.44	1.90	5.500	13.63
6	16¾	1.82	5.50	1.36	4.375	11.25
5	12	1.175	4.60	1.25	4.000	6.37
5	10	.995	4.62	1.01	4.063	5.38
4	10	1.14	3.58	.72	2.900	4.50
4	6	.545	3.65	.71	3.250	2.45

To use the table, compute the maximum bending moment as before explained, and select the beam having the largest value of R nearest to the computed one, in case there is none having the exact required value.

COMPOUND GIRDERS.

For beams compounded from plates and angles, the process for determining R is precisely the same as for any other beam. Inasmuch as compound beams are specially designed for any given case, it is necessary to determine from R the area of the flanges and web, from which the proportions of the parts can be made out. It must be remembered that M or R do not represent *strain*, being independent of *depth*, but can be converted into flange strain by dividing by the depth in inches. Assume, therefore, any depth for the girder (bearing in mind that the effective depth is the distance between centres of gravity of the flanges*), divide R by this depth, and the result is the strain on either flange; and if the maximum allowable strain per square inch has not already been introduced in determining R, the strain above found must be divided by this maximum unit strain to determine the square inches that must be given to the flanges.

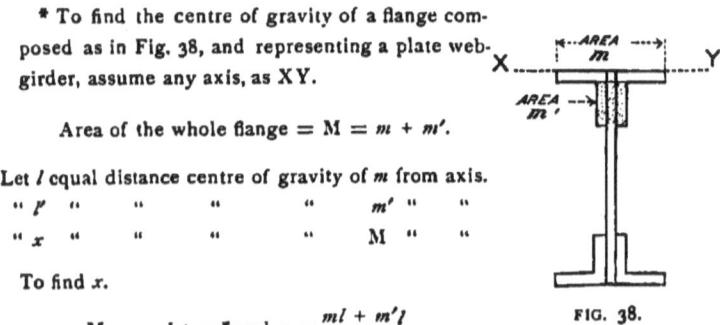

FIG. 38.

* To find the centre of gravity of a flange composed as in Fig. 38, and representing a plate web-girder, assume any axis, as XY.

Area of the whole flange $= M = m + m'$.

Let l equal distance centre of gravity of m from axis.
" l' "   "   "   "   " m' "   "
" x "   "   "   "   " M "   "

To find x.

$$Mx = ml + m'l' \text{ and } x = \frac{ml + m'l'}{M}$$

Taking the same case as before, it is required to know the flange area of a compound beam 15 feet long, 15 inches deep, with a half-inch web. The load being 40,000 lbs. uniformly distributed, the maximum strain to be 10,000 lbs. per square inch, and the assumed effective depth 13 inches. Then, by formula 5, $\frac{40,000 \times 180}{8 \times 10,000 \times 13} = 6.92$ square inches, the required flange area, toward which the web contributes $\frac{1}{6}$ of its area, or $\frac{15 \times \frac{1}{2}}{6} = 1.25$ inches, leaving (6.92 less 1.25) to be built up with angle irons, 5.67 square inches section *net*, after rivet-holes are deducted for the tension-flange—an allowance unnecessary to be made for the compression flange, since that flange is not weakened by the removal of metal, if filled in again as it is with the rivets. Since it is customary to allow a less strain per square inch for compression than for tension, both flanges of plate-girders are usually made alike, the area of the bottom determining that of the top. The allowance for rivet-holes in such forms of plate-girders as are being considered is about 15 per cent, and adding that amount to the net area already found, the gross area of the angle irons must be $6\frac{1}{2}$ square inches, which is given by two angles $\frac{1}{2}$ inch thick, and legs $3\frac{1}{2}$ inches long. To check the effective depth assumed, we have (see foot-note, page 113):
$m = 7'' \times \frac{1}{2}'' = 3.5$ inches; $m' = 3'' \times 1'' = 3$ inches, and $M = 6.5''$
$l = \frac{1}{4}''$; $l' = 2''$.
$x = \frac{3.5 \times 0.25 + 3 \times 2}{6.5} = 0.808$ inches from outer

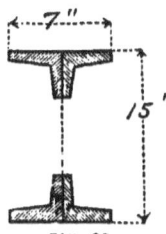

FIG. 39.

edge for each flange, making 1.616 inches as the amount that the full depth is reduced, or 15" less 1.616 inches, equal to 13.38 inches, being practically the same as the effective depth assumed.

In comparing riveted beams with solid rolled beams, it must not be forgotten that the latter are at least 10 per cent stronger than the former; or, in other words, if 10,000 is the unit of strain selected for the riveted work, the solid beam will have as great strength if proportioned for a unit of strain of 11,000 lbs. per square inch.

In order to develop the full strength of a riveted beam, due to the section, more attention should be paid to the riveting than is usually done, as to number, size, pitch, and method of driving. The duty of the rivets is to take up all the horizontal increments of strain delivered by the web to the flanges. The horizontal strains in the flanges diminish in intensity either way from position of maximum M at centre, toward either abutment, where they are least, and may be found at any point by dividing the moment at that point by the effective depth. The horizontal increments of the *web*

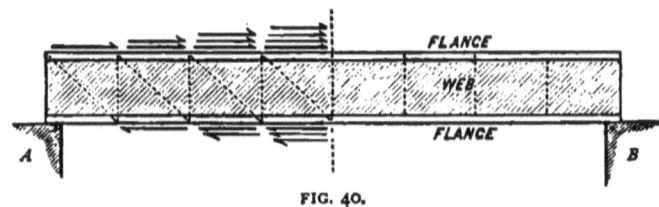

FIG. 40.

are greatest, however, at the ends, and *least* under position of maximum M. This can be made clear from an inspection of the accompanying illustration (Fig. 40), where A B represents a girder loaded uniformly. The web is divided into four imaginary panels, either side of centre, and the horizontal effect of each panel and their summation, as the centre is approached, being represented by arrows. The relative intensity of the horizontal effect is indicated by the varying thickness of the arrows. It has before been stated that the value of a rivet is its diameter multiplied by the thinnest plate through which it passes, multiplied by the working strain per square inch. In the case of a plate-girder, this thinnest plate would be the web, and, assuming such a plate $\frac{3}{8}$ inch thick, a $\frac{7}{8}$ rivet under a working strain of 10,000 lbs. per square inch would have a value of $10{,}000 \times \frac{7}{8} \times \frac{3}{8} = 3300$ lbs. If the maximum horizontal strain is divided by 3300 lbs., there results the minimum number of rivets required either way from the point of such strain. Owing, however, to the greater intensity of the horizontal increments of strain of the web toward the ends, the rivets should be spaced closer as the ends are approached. Applying these rules to the girder previously computed, the loading of which brought the maximum strain at the centre, this strain was found to be 69,200 lbs.; and using a $\frac{3}{8}$-inch web and $\frac{7}{8}$ rivets as above, we find that the number of rivets required either way from the centre to the ends, a distance of 90 inches, will be $\left(\frac{69{,}200}{3300}\right)$ about 21, which, if uniformly pitched, would be spaced a shade over 4 inches between centres. It will be better, however, for the reasons above given, to use more rivets, spacing them 3 inches for the first end quarter, or for 45 inches, the balance being 4½ inches pitch. At first sight, from theoretical considerations purely, it would seem that a good proportioning of riveted work would require a variation in size of rivets, but such designing would cause endless trouble during manufacture. Uniformity of parts in design is essential to economical production, as well as for the avoidance of shop errors, and for this reason, in flange-riveting, the same size rivets should be used throughout, and change of pitch avoided as much as possible. Some manufacturers depend more or less upon the efficiency of rivets being increased by reason of the friction of the rivet-heads against the plates, due to their shrinkage after being driven. There is no doubt but that in new work this friction is very great, and materially aids the rivet, but as it is uncertain how much this is impaired after a long term of service, as well as the variability of the value of the friction, it is deemed by the most prudent designers of iron-work to make no allowance whatever for friction, but proportion rivets only with reference to their bearing surfaces and shearing areas. As to stiffeners for the webs of girders in highway bridges, they are unnecessary if the thickness of the plate is such that the unsupported distance between the legs of the upper and lower flange angle iron is not greater than from 35 to 40 times that thickness. If this proportion is exceeded, stiffeners must be introduced at intervals and over the points of support. Since the floor-girders of a highway bridge are proportioned (or should be) for the extreme standard load, the rarity of such occurrence, if it ever really occurs at all, is such as warrants the recommendation of a unit strain of 15,000 lbs. per square inch to be used for the net section of the lower flange. Angle and plate iron can now be readily obtained of a good quality, with elastic limits up to 23,000 lbs. per square inch, and it is an absurd waste of material to use a low unit strain for the exceptional circumstances of extreme loading. Power-riveting is so superior in all respects to hand-riveting that a higher unit of strain, by probably 10 per cent, can be used under the former system; so that if it is considered proper to strain hand-riveted work up to 13,500 lbs. per square inch, work riveted up by steam or hydraulic power can be safely proportioned on a basis of 15,000 lbs. per square inch.

For convenience of selection of rivets, the following table has been prepared, giving the working values of different sized rivets, in pounds, for plates of varying thickness, and for different units of strain per square inch. Rivet-holes are punched or drilled $\frac{1}{16}$ inch larger than the rivets, and if the holes are properly filled, as they can be by power-riveting, their effective diameter is correspondingly increased.

Size of Rivets.	Thickness of Web Plates.														
	For 10,000 lbs. per sq. inch.					For 12,000 lbs. per sq. inch.					For 15,000 lbs. per sq. inch.				
	¼	⁵⁄₁₆	⅜	⁷⁄₁₆	½	¼	⁵⁄₁₆	⅜	⁷⁄₁₆	½	¼	⁵⁄₁₆	⅜	⁷⁄₁₆	½
	Lbs	Lbs	Lbs	Lbs	Lbs	Lbs	Lbs	Lbs	Lbs	Lbs	Lbs	Lbs	Lbs	Lbs	Lbs
½ inch.	1250	1560				1500	1875				1875	2350			
⅝ "	1560	1950	2340			1875	2350	2810			2350	2930	3510		
¾ "	1875	2340	2810	3280	3750	2250	2810	3380	3940	4500	2810	3510	4220	4920	5630
⅞ "	2190	2810	3280	3830	4375	2630	3290	3940	4600	5250	3280	4100	4920	5740	6560
1 "	2500	3125	3750	4375	5000	3000	3750	4500	5250	6000	3750	4690	5630	6560	7500

STRAINS IN TRUSSES.

In the following discussion of this subject no attempt will be made to go beyond the ordinary forms in constant use, since to do so would be foreign to the object of the writer, as explained in the preface. So many excellent treatises have been written on this subject, that any student desirous of going beyond these elementary pages has a large field to choose from. Probably the best general work on the subject is that of Mr. S. H. Shreve (published by D. Van Nostrand, New-York), inasmuch as the method of analysis therein adopted refers all forms of trussing to the principle of the lever, no special analysis being employed for each case as it arises. The development of a subject from one simple root or principle permits of an intellectual grasp of that subject impossible to attain by the discussion of its separate topics in an independent manner, even if independent analysis were more readily performed. It is not one of the least of the beauties of the method of the lever that, its elementary principles being once mastered, it can be immediately applied to any system of trussing without reference to formulæ, and is therefore an immense relief to the memory. It is believed that in the previous discussion of the strength of beams, the principle of the lever has been so thoroughly kept in view that its application to truss forms will be readily appreciated. Before so applying it, however, it is necessary to explain some elementary ideas of the composition and resolution of *forces*.

The composition of a force is the operation of finding a single force whose effect is equivalent to two or more single forces, while the resolution of a force is the converse operation, being the operation of finding two or more forces the equivalent of a given single force. In mechanics, forces are represented by straight lines, both as to magnitude and direction, by taking the lines proportional to the forces which they represent. What is called the parallelogram of forces is as follows: "If two forces be represented in direction and intensity by the adjacent sides of a parallelogram, their resultant will be represented in direction and intensity by that diagonal of the parallelogram which passes through their point of intersection." Thus in Fig. 41 the two forces are represented by the lines P and Q applied to a material point O.

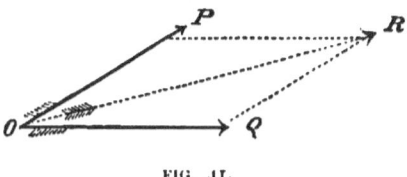

FIG. 41.

Then the same effect will be produced on that point if the two forces are removed and the diagonal R, called the resultant, substituted. If the diagonal force is exerted in the direction of the arrow, motion will result. If in the *contrary* direction, P and Q supposed to be in action as shown, there is rest or equilibrium. It will be noticed that a triangle can be substituted for the parallelogram, by laying off *from* Q a line equivalent and parallel to P, when the resultant is at once obtained by drawing a line from the end of P, thus transferred, to O, and it is in this form that the parallelogram of forces is usually applied. Thus modified, we have what is called the *triangle* of forces, and the law that " If three forces acting at one point balance, three lines parallel to their directions will form a triangle, the sides of which are proportional to the forces, both in direction and intensity." Thus if there are three forces, P, Q, R, Fig. 42, acting at the point O, and if we draw to any scale A B parallel to R; A C and B C parallel to P and Q, the sides of the triangle thus formed will be proportional to the amounts of the forces P, Q, and R. If R is known, then B and C may be scaled off or computed from the geometrical relation of the sides of a right-angled triangle. A C and B C are called the *vertical* and *horizontal components* of A B, or the equivalent *effect* of A B in a vertical or horizontal direction. Any force acting at an angle can be determined, therefore, if we know either component and the angle of inclination from the simple relation of the parts of a triangle.

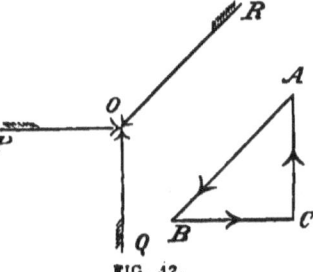

FIG. 42.

Thus far we have spoken of *force*, but in reality we know nothing of force itself, but only the *effects* which it produces. These effects in structures are called *strains*, which, when acting in the direction of the length of any bar or member of a frame, are called longitudinal strains.

Let A B. Fig. 43, be a known vertical effect of a force, and let the geometrical relations of the lines of the triangle be also known. Then the longitudinal strain in A C will exceed the vertical strain in A B by the number of times A B is contained in A C. For example, assume a right-angled triangle, the relations of whose sides are 6, 8, and 10, and suppose the effect of force on A B has been to produce a strain of w lbs., then the longitudinal strain in A C is w lbs. multiplied by $\frac{10}{8}$, or $1\frac{1}{4}$ times the vertical strain. The strain on C B will be similarly $\frac{6}{8}$ times w, or $\frac{3}{4}$ the vertical strain. For a wonderfully clear and elaborate discussion of force, strains, etc., as well as upon the subject of trusses and strength of materials, freed from all technicalities, the learner is referred to Mr. Trautwine's "Engineer's Pocket-Book," a work that should be the corner-stone of every engineer's library.

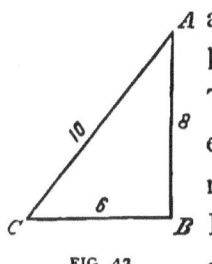

FIG. 43.

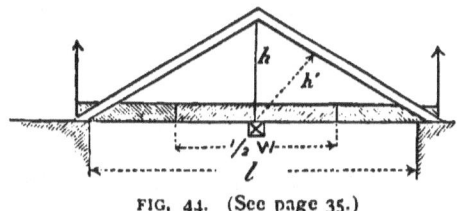

FIG. 44. (See page 35.)

THE KING POST TRUSS.—The extreme effect on all parts of this form of truss occurs when loaded with the combined live and dead loads. In the construction shown in the figure, one half the whole load rests upon the cross-beam upheld by the kingbolt, the other half being carried by the abutments directly, and does not affect the truss at all.

Call the span l; the height, h; load per foot, w, whence total load $= lw = W$. For equilibrium the moment of the external forces must be equal to the moment of the internal forces. The external force is the reaction of the abutment, or $\frac{1}{4} wl$; the internal force is the strain on the material. Taking moments around the *foot* of the kingbolt, we have for the thrust in either rafter: Reaction multiplied by its lever = thrust multiplied by its lever, or $\frac{wl}{4} \times \frac{l}{2} = T \times h'$ and $T = \frac{wl^2}{8h}$.

For the pull on the tie-beam, moments must be taken around the apex of the rafters. Reaction multiplied by lever = pull multiplied by lever, or $\frac{wl}{4} \times \frac{l}{2} = P \times h$ and $P = \frac{wl^2}{8h}$.

The strain on the kingbolt is simply the load upheld by it, or $\frac{1}{2} wl$. If instead of carrying the load on horizontal stringers supported midway by a cross-beam, in turn held up by the kingbolt, as in the example, it is distributed over the tie by numerous transverse beams, then the tie, in *addition* to the pull on it from the thrust of the rafters, must be proportioned as an ordinary beam exposed to a uniformly distributed load of $\frac{1}{2} wl$ for each half of the tie. If this truss is turned upside down, the value of the strains will remain as it was before, only being reversed in kind; that is, the kingbolt will become a post, suffering compression, as does the horizontal chord, and the diagonals undergoing tension will become ties.

THE QUEEN POST TRUSS (Fig. 45).—The load is supposed to be carried from panel-point to panel-point by means of stringers, thus avoiding cross-strain on the horizontal chord.

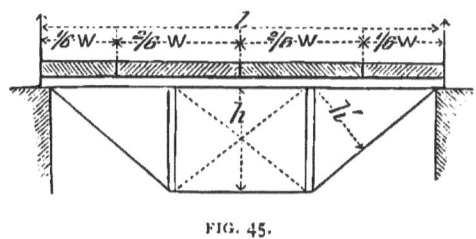

FIG. 45.

Call span l; depth truss, h; w load per ft. = wl, total load; each panel $\frac{1}{3}l$. Excepting on dotted diagonals, maximum strains occur when load is on both posts. Reaction of either abutment will be half the load supported by the truss, or the load on one post = $\frac{1}{3}wl$. For the horizontal strain of compression in the top chord, engendered by the pull of the end diagonals, take centre of moments around the foot of either post. The forces in action are the reaction of either abutment and the strain on the material, the lever-arm of the former being one panel-length, and of the latter the depth of the truss. As these forces must balance, there results $\frac{wl}{3} \times \frac{l}{3} = T \times h$, or thrust = $\frac{wl^2}{9h}$. The pull in the parallel bottom chord will manifestly be of the same amount. The strain in the chords being derived solely from the end diagonals, the strain in the latter may be determined from it by remembering that the chord strain is the horizontal effect or the component of either diagonal, of which the vertical effect or component is the load on the post, or the abutment reaction (page 122). Knowing then the horizontal component, the longitudinal strain in the diagonal is given by multiplying this component by the length of the diagonal, and dividing by the length of the panel; or

Tension strain in diagonal = Hor. thrust $\times \frac{\text{length of diagonal}}{\text{length of panel}}$.

The compression on the post is simply the panel load upon it, and equals, therefore, $\frac{1}{3} wl$.

The dotted diagonals are counter-braces, and are only brought into play when but *one* post is loaded, in which case the reaction of the abutment nearest the load is twice as great as that of the other. This difference of reactions yields unbalanced horizontal components for the main diagonals, which must be counteracted by a tension-bar, the horizontal component of which must equal the difference between the horizontal components of the main diagonals. Thus, suppose the load of $\frac{1}{3} lw$ on the right-hand post is removed; the reaction of the right abutment will be, according to the law of the lever, $\frac{1}{3}$ of this, or $\frac{1}{9} lw$, and that of the left abutment will be $\frac{2}{9}$ of the panel load, or $\frac{2}{9} lw$. The horizontal component, or the chord strain from the left diagonal, is

$\frac{2}{9} lw \times \frac{1}{3} l$ divided by $h = \frac{2 l^2 w}{27 \times h}$, and for the right diagonal this component is $\frac{1}{9} lw \times \frac{1}{3} l$ divided by $h = \frac{l^2 w}{27 \times h}$.

The difference between these two values is $\frac{1}{27}$ of $\frac{l^2 w}{h}$, and is the horizontal component of the counter diagonal sought. Its longitudinal strain is found as before for the main diagonal braces, by multiplying the component just found by its length, $\sqrt{\frac{1}{9}l^2 + h^2}$, and dividing the product by the panel-length, or $\frac{1}{3}l$; or,

Maximum counter-strain of tension $= \frac{l^2 w}{27 h} \times \sqrt{\frac{\frac{1}{9}l^2 + h^2}{\frac{1}{3}l}}$.

If the load is distributed over the whole upper chord instead of being concentrated at panel-points, to the longitudinal thrust, due to its position in the truss, must be added the requirements of a beam uniformly loaded. As the Queen Post truss is the parent of the most usual forms of truss met with, the following numerical example is given of the application of the preceding principles:

Data.—$l = 45$ ft.; $\frac{1}{3}l =$ panel-length $= 15$ ft.; truss 5 ft. deep.

$w = 300$ lbs. per ft. $= 4500$ lbs. per panel $=$ dead load $\Big\}$
$w' = 2000$ " " $= 30,000$ " $=$ live load $\Big\}$ $= 34,500$ lbs. total panel load.

Length of diagonal $= 15.81$ ft. Abutment reaction when wholly loaded, 34,500 lbs.

Strain on horizontal chords $= T$

$\frac{wl^2}{9h} = \frac{2300 \times 45^2}{9 \times 15} = 103,500$ lbs. compression upper and tension lower.

Strain on post—

$\frac{1}{3} wl = \frac{1}{3} 2300 \times 45 = 34,500$ lbs. compression.

Strain on end diagonals—

$T \times \sqrt{\frac{\frac{1}{9}l^2 + h^2}{\frac{1}{3}l}} = 103,500 \times \frac{15.81}{15} = 109,089$ tension.

THE WHIPPLE TRUSS.

Strain in counter-braces, one post unloaded. In this case, as the dead load is unchangeable, we are concerned with the live load alone, or 2000 lbs. per ft. = 30,000 lbs. per panel. The reaction of the left abutment from this (supposing the post to the right is unloaded) is 20,000 lbs., and of the right abutment 10,000 lbs.

$$\left.\begin{array}{l}\text{Horizontal strain from left diagonal, } \frac{20,000 \times 15}{5} = 60,000 \\ \text{Horizontal strain from right diagonal, } \frac{10,000 \times 15}{5} = 30,000\end{array}\right\} \text{diff. } 30,000.$$

This difference being horizontal difference, for conversion into longitudinal strain on the counters, it is to be multiplied as before by $\frac{15.81}{15}$, which gives 31,620 lbs. as tension on the counters; or by applying the formula, the strain is at once given—

$$\frac{l^2 w}{27 \times h} \times \sqrt{\frac{\frac{1}{4} l^2 + h^2}{\frac{1}{4} l}} = \frac{45^2 \times 2000}{27 \times 5} \times \frac{15.81}{15} = 30,000 \times 1.054 = 31,620, \text{ as before.}$$

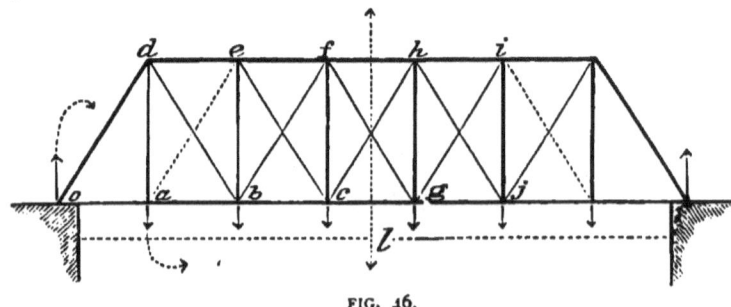

FIG. 46.

THE WHIPPLE TRUSS (Fig. 46).—By extending the Queen Post so as to embrace additional panels, the

Whipple truss is developed, as in the figure representing the diagram of a through-bridge having seven panels.

Let l = span; n = number panels; h = height of truss; w dead load at each panel-point; w' = variable load on one panel.

1st. Chord Strains.—Maximum strain in chords occurs when all panels are loaded with dead and live loads, in which case reaction of either abutment is $\frac{(w + w')(n-1)}{2}$, or three panel loads = $3(w + w')$. For first panel, horizontal strain will be (moments around d as a fulcrum) $3(w + w') \times \frac{l}{n} \div h$, or reaction multiplied by lever of one panel-length, divided by depth of truss. For third panel bc, the strain will be (moments around b) $\left(3(w + w') \times \frac{2l}{n} - (w + w') \times \frac{l}{n}\right) \div h$. In this expression it will be noticed that one panel load multiplied by its lever of one panel is *subtracted* from the moment of the reaction. This is because the weight at a operates downward or contrary to the reaction of the abutment, as shown by dotted lines, and reduces correspondingly the effect of abutment reaction. On the middle panel cg, the horizontal strain will be $\left(3(w + w') \times \frac{3l}{n} - (w + w' \frac{2l}{n}) - (w + w')\frac{l}{n}\right) \div h$; or, in other words, subtract from the moment of the reaction—operating in one direction—the moments of the panel loads between fulcrum and abutment, and then divide by the depth for the strain. The same process must be continued for any number of panels up to the centre of the truss where the strains are a maximum, after which they decrease to the other abutment. While the chord strains are the same top and bottom, they are not so for the same panel. The inclination of the diagonals brings the panels of equal strain in advance of each other; that is to say, the tension strain in $b\ c$ is the same as the compressive strain in $d\ e, c\ g$ as in $e\ f$. The example given is for an uneven number of panels, in which case there will be three panels of the top chord—namely, the centre and one either side, of equal maximum strain, to one panel of maximum strain at the centre of bottom chord. If a diagram is made for an even *number* of panels, there will be a post at the centre, and it will be seen that the maximum strain on top chord will extend over two panels, one on either side of centre, and will be in excess of the maximum strain in the bottom chord, owing to the main diagonals of the two middle panels *uniting* at the foot of the post where their horizontal components balance each other. At the top chord, however, these diagonals are spread apart two panel-lengths, and deliver their horizontal component to that chord.

Example.—Let $l = 70$ ft.; $n = 7$; $h = 10$ ft.; $w = 300$ lbs. ft. $= 3000$ lbs. per panel; $w' = 1000$ lbs. ft. $= 10,000$ lbs. per panel; $w + w' = 13,000$; abutment reaction $=$ one half of six panel loads $= 39,000$ lbs.

The horizontal strain on O a and $a\ b$ will be
$$\frac{39,000 \times 10}{10} = 39,000 \text{ lbs.}$$

The horizontal strain on $d\ e$ and $b\ c$ will be
$$\frac{39,000 \times 20 - 13,000 \times 10}{10} = 65,000 \text{ lbs.}$$

The horizontal strain on $e\ f, f\ h, h\ i,$ and $c\ g$ will be
$$\frac{39,000 \times 30 - 13,000 \times 20 - 13,000 \times 10}{10} = 78,000 \text{ lbs.}$$

2d. *Web Strains.*—The web strains must be computed separately under each condition of loading. The posts and braces are strained the greatest when the moving load covers the segment *from* which any given diagonal slopes. Thus the diagonal $e\,c$ is strained the greatest when c and all points to the right are loaded with moving load; $f\,g$ when g and all points to its right are loaded, etc. While the web strains can be readily calculated by finding the horizontal components for each maximum condition of loading, and converting them into longitudinal strains, as was done for the Queen Post truss, the method is somewhat tedious when there are a number of panels, and a separation of dead and live loads must be made. For trusses with parallel chords, the following method will be found most convenient, and is the one usually employed. It is based on considering the load on each panel-point, tracing its action on the posts and ties, and summing their effects—or, in other words, finding the vertical components, which are the post strains. Taking first the dead load, there is w at each panel-point, or, under the example, 3000 lbs. Since three panel loads are supported by each abutment, the loads, and therefore the strains, are symmetrical with the centre, and it is only necessary to compute the strains for one half the truss. At the point c, 3000 lbs. is taken up by the inclined tie $e\,c$, and delivered to the vertical post $e\,b$, which has a compression, therefore, of that amount; the tension on the tie being 3000 lbs. $\times \frac{\text{its length}}{\text{depth of truss}}$, or $3000 \times \frac{14.1 \text{ ft.}}{10 \text{ ft.}} = 4230$ lbs. This panel load is again progressed to the abutment by the tie $d\,b$, which also has upon it another panel load at b of 3000, making 6000 lbs. delivered to the end-post $d\,o$. The strain on this tie is, therefore, just double that on the preceding tie, or 8460 lbs., to which must be added the effect of the third panel load sustained by the vertical tie $d\,a$, or 4230 lbs. for the compressive strain for the inclined end-post, making a total for that post of 12,690 lbs. For the moving load alone, advancing from the left abutment, we have, when it reaches the point a, 10,000 lbs. By the law of the lever, $\frac{5}{6}$ of this is supported by the left abutment, and $\frac{1}{6}$ by the right abutment. Since the whole of this load ascends the vertical $a\,d$, the $\frac{1}{6}$ that goes to the right can only do so by passing down the diagonal $d\,b$ to the foot of the post $e\,b$, when the diagonals in the opposite direction progress it toward the right abutment. The strain in $d\,b$ from this action of the load is one of compression; but since the dead load strains this diagonal *tensively* largely in excess of this compressive effect, the latter is entirely neutralized. Advancing to each panel-point in succession with the load of 10,000 lbs., and distributing the load by the law of the lever, the strains on the various parts will be as follows, from the live load alone:

On—$o\,d$: $\frac{1}{6}(6+5+4+3+2+1)$ 10,000 $\times \frac{14.1}{10}$ compression = 42,300
$a\,d$: one panel load, tension = 10,000
$d\,b$: $\frac{1}{6}(5+4+3+2+1)$ 10,000 $\times \frac{14.1}{10}$ " = 30,214
$e\,b$: $\frac{1}{6}(4+3+2+1)$ 10,000 compression = 14,280
$e\,c$: same as $e\,b \times \frac{14.1}{10}$ tension = 20,143

fc: $\frac{1}{4}(3+2+1) \times 10{,}000$ compression = 8,571
fg: same as $fc \times \frac{14}{10}$ tension   = 12,086
hg: same maximum as fc compression = 8,571
hj: $\frac{1}{4}(2+1)\, 10{,}000 \times \frac{1414}{10{,}000}$ tension   = 6,040

If to these are added the previously computed effects of the dead load, there result the maximum strains that can come upon the web system by any possible condition of loading.

Since the counter diagonals can only act when the main diagonals of the same panel are relaxed, it follows that to obtain the maximum tension of any counter, the effect of the dead load to which it is opposed must be subtracted from the effect due to the live load alone. Thus, counter hj is strained from the live load 6040 lbs., but main diagonal gi is strained by the dead load 4230 lbs; therefore the counter is to be proportioned for 6040 less 4230 lbs., or only 1810 lbs.

In the case of the Whipple double-cancelled truss, each system of the web must be computed independent of the other, and their joint effect on the chords added.

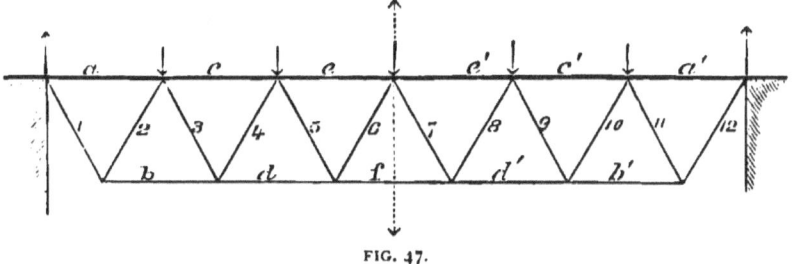

FIG. 47.

WARREN GIRDER (Fig. 47).—Load supposed to be concentrated at the panel-points of the top chord.

Span, 60 feet; 6 panels of 10 feet. Truss, 10 feet depth. Length diagonal, 11.18 feet. Dead load per panel, 3000 lbs. Moving load per panel, 9000 lbs. Reaction of abutment for full loading $\left(\frac{9000+3000}{2}\right) \times 5$ panels = 30,000 lbs.

For Chord Strains, the centre of moments must be taken alternately on top and bottom at panel-points. It will be noticed that at each panel-point, owing to the inclination of the web members, there is a horizontal effect from each member, acting in the same direction.

Compression on a, $\dfrac{30{,}000 \text{ lbs.} \times 5 \text{ ft.}}{10 \text{ ft.}} = \ldots$ 15,000 lbs.

Tension on b, $\dfrac{30{,}000 \text{ lbs.} \times 10 \text{ ft.}}{10 \text{ ft.}} = \ldots$ 30,000 "

Compression on c, $\dfrac{30{,}000 \text{ lbs.} \times 15 \text{ ft.} - 12{,}000 \times 5}{10} =$ 39,000 "

Tension on d, $\dfrac{30{,}000 \times 20 - 12{,}000 \times 10}{10} = \ldots$ 48,000 "

Compression on e, $\dfrac{30{,}000 \times 25 - 12{,}000 \times 5 - 12{,}000 \times 15}{10} =$ 51,000 "

Tension on f, $\dfrac{30{,}000 \times 30 - 12{,}000 \times 10 - 12{,}000 \times 20}{10} =$ 54,000 "

The preceding operation is simply putting in figures the oft-repeated principle of the lever. In each case the numerators of the fractions may be read, "The reaction of abutment (upward) multiplied by its lever, less the separate panel loads (downward) between abutments and fulcrum, multiplied by their levers;" while the denominator is the depth of truss, which is the leverage of resistance.

FOR WEB STRAINS.—1st. *Dead-load.* One half of that at central apex, or 1500 lbs., goes down diagonal 6, up 5, down 4, which receives in addition a full panel-load, making 3000 + 1500 lbs., which goes up diagonal 3, being again increased before passing down 2, with another panel-load, or 3000 + 3000 + 1500 lbs., which, in turn, passes up diagonal 1 to point of support. The diagonals to right of centre are traversed by the load on that side in the same way. These vertical effects need only to be multiplied by $\frac{\text{length of diagonal}}{\text{height}}$ to give the sought-for longitudinal strains in the diagonals due to dead-load. When the load passes down, *compression* is induced; and when up, *tension.* Thus, 1, 3, and 5 are in tension, and 6, 4, and 2 are in compression. 2d. *Variable load.* Commence by loading the first apex on the left with moving panel-load 9000 lbs.; of this $\frac{5}{6}$ is supported by the left abutment, and $\frac{1}{6}$ by the right abutment. These proportions of the load only reach their destination by passing down and up alternately the different web members, inducing compression and tension alternately. Tracing out the effect of each load (p, q, r, s, t) in succession, commencing at the left apex, the vertical effect on diagonals will be as follows:

$\frac{5}{6} p + \frac{4}{6} q + \frac{3}{6} r + \frac{2}{6} s + \frac{1}{6} t$ all produce tension on 1.
$\frac{5}{6} p + \frac{4}{6} q + \frac{3}{6} r + \frac{2}{6} s + \frac{1}{6} t$ "   "   compression on 2.
3 receives a compression from $\frac{1}{6}$ the load at p, and tension from all loads to the right, amounting to $\frac{4}{6} q + \frac{3}{6} r + \frac{2}{6} s + \frac{1}{6} t$.

STRAINS IN THE WARREN GIRDER. 135

Diagonal 4 has tension from $\frac{1}{6}p$, and a compression from $\frac{2}{6}q + \frac{3}{6}r + \frac{2}{6}s + \frac{1}{6}t$.
" 5 " compression from $\frac{1}{6}p + \frac{2}{6}q$, and a tension from $\frac{3}{6}r + \frac{2}{6}s + \frac{1}{6}t$.
" 6 " tension from $\frac{1}{6}p + \frac{2}{6}q$, and a compression from $\frac{3}{6}r + \frac{2}{6}s + \frac{1}{6}t$
" 7 " compression from $\frac{1}{6}p + \frac{2}{6}q + \frac{3}{6}r$, and a tension from $\frac{2}{6}s + \frac{1}{6}t$.
" 8 " tension from $\frac{1}{6}p + \frac{2}{6}q + \frac{3}{6}r$, and a compression from $\frac{2}{6}s + \frac{1}{6}t$.
" 9 " compression from $\frac{1}{6}p + \frac{2}{6}q + \frac{3}{6}r + \frac{4}{6}s$, and a tension from $\frac{1}{6}t$.
" 10 " tension from $\frac{1}{6}p + \frac{2}{6}q + \frac{3}{6}r + \frac{4}{6}s$, and a compression from $\frac{1}{6}t$.
" 11 " compression from $\frac{1}{6}p + \frac{2}{6}q + \frac{3}{6}r + \frac{4}{6}s + \frac{5}{6}t$, no tension.
" 12 " same as 11, only tension.

Summing these effects of the moving load, and remembering that the loads at each apex are the same in amount, or 9000 lbs., $\frac{1}{6}$ of which is 1500 lbs., which, converted into longitudinal effect, is $1500 \times \frac{11.18 \text{ ft.}}{10 \text{ ft.}} = 1677$ lbs., we have for the strains in the web:

Diagonal 1. $15 \times 1677 = 25{,}155$ lbs. tension.
" 2. $15 \times 1677 = 25{,}155$ lbs. compression.
" 3. $10 \times 1677 = 16{,}770$ lbs. tension, and $1 \times 1677 = 1677$ lbs. comp.
" 4. $10 \times 1677 = 16{,}770$ lbs. comp., and $1 \times 1677 = 1677$ lbs. tension.
" 5. $6 \times 1677 = 10{,}062$ lbs. tension, and $3 \times 1677 = 5031$ lbs. comp.
" 6. $6 \times 1677 = 10{,}062$ lbs. comp., and $3 \times 1677 = 5931$ lbs. tension.
" 7. $6 \times 1677 = 10{,}062$ lbs. comp., and $3 \times 1677 = 5931$ lbs. tension.
" 8. $6 \times 1677 = 10{,}062$ lbs. tension, and $3 \times 1677 = 5931$ lbs. comp.
" 9. $10 \times 1677 = 16{,}770$ lbs. comp., and $1 \times 1677 = 1677$ lbs. tension.
" 10. $10 \times 1677 = 16{,}770$ lbs. tension, and $1 \times 1677 = 1677$ lbs. comp.
" 11. $15 \times 1677 = 25{,}155$ lbs. comp.
" 12. $15 \times 1677 = 25{,}155$ lbs. tension.

To the above values must be added, for final maximum web-strains, the effect of the permanent load, 3000 lbs., at each apex, which, converted into longitudinal effect as above, is 3354 lbs. This is done in the following table:

Name of Diago- nal.	From Moving Load Alone.		From Dead Load Alone.		Algebraic Sum of Moving and Dead Load.	
	+ Com- pression.	− Tension.	+ Com- pression.	− Tension.	+ Com- pression.	− Tension.
	lbs.	lbs.	lbs.	lbs.	lbs.	lbs.
1	....	25,155	....	8,385	....	33,540
2	25,155	....	8,385	....	33,540	none
3	1,677	16,770	....	5,031	none	21,801
4	16,770	1,677	5,031	....	21,801	none
5	5,031	10,062	....	1,677	3,354	11,739
6	10,062	5,031	1,677	....	11,739	3,354
7	10,062	5,031	1,677	....	11,739	3,354
8	5,031	10,062	....	1,677	3,354	11,739
9	16,770	1,677	5,031	....	21,801	none
10	1,677	16,770	....	5,031	none	21,801
11	25,155	....	8,385	....	33,540	none
12	....	25,155	....	8,385	....	33,540

It will be seen from the above table how the compression due to the variable load in diagonals 10 and 3 is more than neutralized by the tension from the fixed load. Diagonals 5, 6, 7, and 8, however, must be capable of acting either by tension or compression, since the effect of the variable load preponderates over the dead load that works against it. In other words, the necessary counterbracing is confined to the last diagonals named. When the span becomes so great as to make the triangles of the Warren system too large, another series may be introduced, each one being computed independently of the other, care being taken not to omit their joint effect on the chords. By increasing the number of systems of triangles, the lattice-truss is formed; but this is not a commendable form of truss, since the intersections of the different systems must be riveted together, which vitiates more or less the calculations, based, as they necessarily must be, upon the hypothesis of an independent action of each system of triangles.

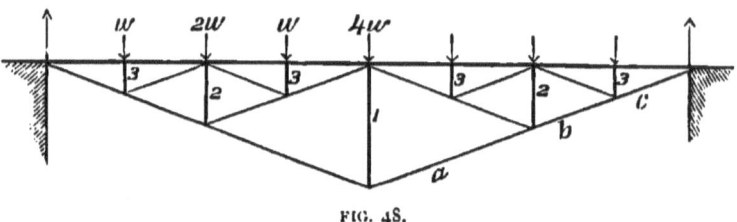

FIG. 48.

THE FINK SUSPENSION TRUSS (Fig. 48).—This form of truss is only well adapted for deck spans, and is precisely the same truss as the ordinary iron roof turned upside down, with the reversal of the quality of strains. The maximum chord strains and all parts of the primary system (marked 1) occur when all points are loaded. On the secondary system (marked 2), maximum strains occur when all the panels embraced in that system alone are loaded, and so on. Let the load on each apex be w, then posts 3 support w only; posts 2 support $w + \frac{1}{2} w$, delivered to it from each sub-post 3, or $2w$; post 1 sustains its own w, $+ \frac{1}{2}$ of the load on each sub-post 2, $+ \frac{1}{2}$ the load from each adjoining sub-post 3, in all $4w$. The suspension rods are strained in proportion to their inclination or $\frac{\text{length of rod}}{\text{height of post}}$. *Example.*—Span, 120 feet; 8 panels, 15 feet; height of centre-post, 15 feet; load on each apex, 10,000 lbs.; ratio of length of any rod to post of system to which it belongs, $\frac{61.8 \text{ ft.}}{15 \text{ ft.}} = 4.12$; ratio of horizontal to vertical, $\frac{60}{15} = 4$.

1st. Strain on posts—compression.

Sub-system, $3 - w = 10,000$ lbs.; secondary, $2 - 2w = 20,000$ lbs.; primary, $1 - 4w = 40,000$ lbs.

2d. Longitudinal tension in suspension bars.

Suspension bars, sub-system 1 . . . $\frac{1}{2}$ 10,000 × 4.12 = 20,600 lbs.
" " secondary system 2 . $\frac{1}{2}$ 20,000 × 4.12 = 41,200 "
" " primary " 3 . $\frac{1}{2}$ 40,000 × 4.12 = 82,400 "

Strain in panel a will be, therefore, 82,400 lbs.; in $b = 82,400 + 41,200 = 123,600$; in $c = 123,600 + 20,600 = 144,200$ lbs. The horizontal chord strain will be *uniform throughout*, and is the sum of the horizontal components of the several systems.

From sub-system 3 . . . . . . $\frac{1}{2}$ 10,000 × 4 = 20,000 lbs.
" secondary system 2 . . . . $\frac{1}{2}$ 20,000 × 4 = 40,000 "
" primary " 1 . . . . $\frac{1}{2}$ 40,000 × 4 = 80,000 "

Total chord strain, ... 140,000 lbs.

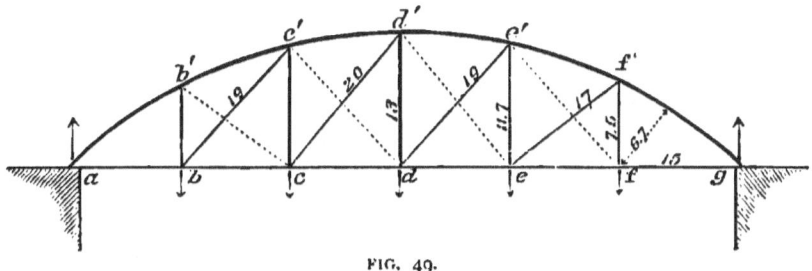

FIG. 49.

THE BOWSTRING TRUSS (Fig. 49).—The maximum horizontal strain occurs when all panels are loaded both with fixed and moving loads, and is *uniform* throughout the length of the tie or bottom chord. The longitudinal thrust through the arch varies with the inclination of the arch-segments, being equal in amount to that of the horizontal strain at the centre only. To find the horizontal strain at the centre under uniform load, "multiply the abutment reaction (in this case $2\frac{1}{2}$ panel-loads) by its lever or $\frac{1}{2}$ span, from which subtract the intermediate panel-loads, multiplied by their leverages, acting in the opposite direction to the reaction, and divide the result by depth of truss." The extreme longitudinal thrust in the arch occurs in the last segment, being the one of greatest inclination, and is at once found by "multiplying the reaction by the lever of one panel-length, and dividing by the perpendicular let fall from the point around which the moments are taken upon the direction of the segment." Or the longitudinal strain in any segment may be found by multiplying the maximum horizontal strain by the length of segment, and dividing by its horizontal stretch.

In the web, under uniform loading, there is no other strain than tension on the verticals, amounting to a panel-load, and the diagonals are unnecessary; but under a variable load, moving from end to end of the truss, the verticals are brought under a compressive strain through the medium of the diagonals, the strain on which may be most conveniently computed as follows: For each position of the load as it advances from point to point, determine the abutment reaction as for an ordinary truss on the principle of the lever. From this compute the horizontal strain at the extreme point of loading, and also at the next panel-point beyond. The difference between these two strains will be the horizontal component of the diagonal of the panel between the points where the horizontal was computed. This has now to be converted into the direction of the diagonal for its longitudinal strain, from which the vertical effect of compression on the post is readily derived. Since tension forever exists on the verticals from the dead load, the amount of tension of one panel dead load must be deducted from the compression above found for maximum compressive effect that can come on a post. As an example of the application of these principles, assume a bowstring truss, with 6 panels of 15 feet, and 13 feet deep at centre. Also let dead load $w = 5000$ lbs. per panel, and live load $w' = 15,000$ lbs. per panel. The lengths of the verticals and diagonals as marked on the diagram:

Maximum horizontal chord strain

$$\frac{\overbrace{(w+w') \times 2\frac{1}{2} \text{ panels} \times 45 \text{ ft.}}^{\text{Reaction.}} l. - (w+w') \times (1+2) 15 \text{ ft.}}{13} = 103,846 \text{ lbs.}$$

Maximum thrust in last segment fg of arch =

$$\frac{(w+w') \times 2\frac{1}{2} \text{ panels} \times 15 \text{ ft.}}{6.7 \text{ ft.}} = \frac{20,000 \times 2\frac{1}{2} + 15}{6.7} = 111,940 \text{ lbs.}$$

Maximum tension on verticals $w + w' = 20,000$ lbs.

Constant tension from dead load alone $w = 5,000$ lbs.

Maximum tension on bc' occurs when variable load is at b alone; reaction left abutment $= \frac{5}{6} w' = 12,500$.

Horizontal tension at $b = \frac{12,500 \times 15}{bb'} = \frac{187,500}{7.5} = 25,000$ lbs.

Horizontal tension at $c = \dfrac{12,500 \times 30 - 15,000 \times 15}{cc'} = \dfrac{150,000}{11.7}$

$= 12,820$ lbs.

25,000 less 12,820 lbs. = 12,180 lbs. the horizontal component.

$12,180 \times \frac{18}{18} =$ longitudinal tension in $bc' = 15,328$.

Maximum tension on cd', moving load at b and c.

Reaction left abutment $= \frac{5}{6}w' + \frac{4}{6}w' = 22,500$ lbs.

Horizontal strain at $c \ldots \dfrac{22,500 \times 30 - 15,000 \times 15}{11.7} = 38,460$.

Horizontal strain at $d \ldots \dfrac{22,500 \times 45 - 15,000 \times 15(1+2)}{13} = 25,900$.

38,460 less 25,900 = 12,560 lbs. = horizontal component of cd'.

$12,560 \times \frac{20}{13} =$ longitudinal tension in cd' 16,713 lbs.

Maximum tension on de', moving load at b, c, and d.

Reaction left abutment $\dfrac{5+4+3}{6} w' = 30,000$.

Horizontal strain at $d = \dfrac{30,000 \times 45 - 15,000 \times 15(1+2)}{13} = 51,923$.

Horizontal strain at $e = \dfrac{30,000 \times 60 - 15,000 \times 15(1+2+3)}{11.7}$

$= 38,461$.

51,923 less 38,461 = 13,462 lbs., horizontal component.

$13,462 \times \frac{18}{18} =$ longitudinal tension, 17,052 lbs.

Maximum tension on ef'; all points but f loaded with w'.

Reaction left abutment $\dfrac{5+4+3+2}{6} w' = 35,000$.

Horizontal strain at $e = \dfrac{35,000 \times 60 - 15,000 \times 15(1+2+3)}{11.7}$

$= 64,103$.

Horizontal strain at $f = \dfrac{35{,}000 \times 75 - 15{,}000 \times 15(1+2+3+4)}{7.5}$
$= 50{,}000$.

64,103 less 50,000 = 14,103 = horizontal component.

14,103 × $\frac{17}{13}$ = longitudinal tension in ef, 15,983.

The compressive strain in verticals from a moving load occurs when all panel-points between any given one and the abutment are loaded. Thus $d\,d'$ is compressed the greatest when b and c or e and f are loaded. The strain (supposing the load is at b and c) on $d\,d'$ will be the vertical component from $d\,c'$, less the tension of one panel of dead load. It is necessary, then, to find the longitudinal strain on the different diagonals when the panel-points beyond are loaded, and that of the given diagonal unloaded.

On $c\,d'$, when b alone is loaded, reaction = $\frac{5}{6} w' =$ 12,500.

Horizontal strain at $c = \dfrac{12{,}500 \times 30 - 15{,}000 \times 15}{11.7} = 12{,}820$.

" " $d = \dfrac{12{,}500 \times 45 - 15{,}000 \times 30}{13} = 8654$.

12,820 less 8654 = 4166 = horizontal component, which multiplied by $\frac{20}{13}$ = 5521 = longitudinal strain. Converting this last strain into vertical strain by multiplying it by the ratio of diagonal to vertical, or $\frac{13}{20}$, the compression on post $c\,c'$ from line load is obtained. Since there is always a tension caused by one panel of dead load, the compression above found must be reduced by that amount, to obtain the maximum compression.

THE BOWSTRING TRUSS. 143

On $d\,d'$, when b and c are loaded, reaction $= \frac{9}{8} w' = 22,500$.

Horizontal strain at $d = \frac{22,500 \times 45 - 15,000 \times 15 (1 + 2)}{13} = 25,963$.

Horizontal strain at $e = \frac{22,500 \times 60 - 15,000 \times 15 (2 + 3)}{11.7} = 19,145$.

25,963 less 19,145 = 6818 = horizontal component. Multiplying this component by $\frac{13}{15} = 5900$; less 5000 = maximum compression on post $d\,d' = 900$.

On $e\,e'$, when b, c, and d are loaded, reaction $= \frac{12}{8} w' = 30,000$ lbs.

Horizontal strain at

$$e = \frac{30,000 \times 60 - 15,000 \times 15 (1 + 2 + 3)}{11.7} = 38,461.$$

Horizontal strain at

$$f = \frac{30,000 \times 75 - 15,000 \times 15 (2 + 3 + 4)}{7.5} = 30,000.$$

38,461 less 30,000 = 8461 horizontal component. Converting this horizontal strain into vertical, there results for compression on posts from live load 8461 lbs. $\times \frac{7.5}{15} = 4230$ lbs. Since the tension induced by dead load is 5000 lbs., there can, therefore, be no compression whatever on post $e\,e'$.

On $b\,b'$ or $f\,f'$, there can be no other strain than that of tension from $w + w'$.

If the bowstring is inverted, the strains may be computed in the same way as above explained, but are reversed in quality. The horizontal tie will become a compression chord, and the arch will be under tension.

The posts, in this case, will be compressed from the dead load, the effect of which is therefore *added* to that of the diagonals (being of the same quality), instead of being subtracted as before.

For a deck span this adaptation of the bowstring truss is to be commended as economical in material and pleasing in appearance.

www.ingramcontent.com/pod-product-compliance
Lightning Source LLC
Chambersburg PA
CBHW030310170426
43202CB00009B/951